development efforts. Understanding the different needs and priorities of men and women in Third World countries will enable policy-makers, planners and administrators to improve the impact of their programmes and projects. An indispensable aid to all those who work in development agencies, the book will also be of great interest to students of development studies and women's studies.

Gender and development

Gender relations – the social, economic, political, and legal roles of men and women within a society – vary greatly from culture to culture. These roles have a marked effect on how individuals behave and react to new ideas. The contributors to this book believe that each development effort should be preceded and accompanied by a gender aware analysis which takes into account the needs and roles of males and females in the area where each project will operate. Development programmes and projects which are planned and administered with little insight into gender relations usually fail to have their intended effect.

Recognized experts in their fields, the contributors deal here with the importance of gender relations in the crucial development areas of agriculture, employment, housing, transport, health and management of household resources. Emphasizing the need for statistical methods which reflect realistic gender differentials, they offer new methods and sources for obtaining more reliable information. The final chapter provides a cogent analysis of what is required to incorporate the gender perspective into development efforts, and gives invaluable practical guidelines on how best to instil gender awareness into the administration and activities of donor countries and agencies.

Gender and Development discusses its themes in a very clear, practical manner, showing readers how to involve women more effectively in development efforts. Understanding the different needs and priorities of

Gender and development

A practical guide

Edited by
Lise Østergaard

Based on a study prepared for
the Directorate-General for Development
Commission of the European Communities

London and New York

First published 1992
by Routledge
11 New Fetter Lane, London EC4P 4EE

Simultaneously published in the USA and Canada
by Routledge
29 West 35th Street, New York, NY 10001

Reprinted 1994

Typeset in Times by LaserScript Limited, Mitcham, Surrey
Printed and bound in Great Britain by
Mackays of Chatham PLC, Chatham, Kent

British Library Cataloguing in Publication Data

A catalogue record for this book is available from the British Library.

Library of Congress Cataloging in Publication Data

Gender and development : a practical guide / edited by Lise Østergaard.
p. cm.
Includes bibliographical references and index.
1. Women in development – Developing countries. 2. Sex role – Developing countries. 3. Economic development projects – Developing countries. I. Østergaard, Lise.
HQ1240.5.D44G46 1992
305.42'09172'4–dc20

91–16824
CIP

ISBN 0–415–07131–3 (hbk)
ISBN 0–415–07132–1 (pbk)

Contents

Contributors

Cecilia Andersen is Professor at the Center of Business Administration at the University of Antwerp (University Faculty St Ignatius). She also teaches at Boston University, Brussels. She has published extensively on WID in development cooperation and serves as a consultant for national as well as international organizations.

Helen Bloom is an investigative reporter, writer and editor who contributes her time to developmental issues which concern her. At present she is Executive Director of Belmont European Policy Centre, a Brussels-based think-tank which is known for its expertise on European Community affairs.

Alison Evans is an economist who spent four years working at the Institute of Development Studies. She is now teaching and researching at the School of African and Asian Studies, University of Sussex. She has been teaching and training in gender and development since 1984, specifically on issues in household economics, labour markets and the distributional implications of economic adjustment in developing countries. Her field research has taken her to Zambia and to Uganda, where she is currently working on an institutional economics approach to the study of village labour markets and agricultural development.

Caren Levy is a lecturer with an economics and planning background at the Development Planning Unit, University College London. She is Co-Director of the DPU Gender Planning Programme, which encompasses education, research, training and advisory services in different parts of the world.

Caroline Moser is a social anthropologist/social planner. She is currently Senior Urban Social Policy Specialist in the Urban Development Division, World Bank, Washington DC. Her research interests include gender planning with special reference to employment, housing and basic services, community participation, and the social costs of adjustment policy and economic reform. She is co-editor of *Women, Human Settlements and Housing*.

Lise Østergaard is Professor of Human Studies in International Development at the University of Copenhagen. She also holds consultantships for a number of international organizations concerned with issues on gender and development. Before her present appointment in 1984 she was a member of the Danish Parliament from 1979–84 and Minister for Overseas Development, Minister of Culture, and Minister for Nordic Affairs. She presided over the World Conference of the UN Decade for Women in 1980.

Hilary Standing teaches Social Anthropology at the School of African and Asian Studies, University of Sussex. Her research interests are in social and economic transformation and their effects on gender relations, especially in the context of family and kinship. She has carried out research in both rural and urban India.

Ann Whitehead has had a long teaching and research history in issues of gender and development, and specializes in sub-Saharan Africa. She trained as a social anthropologist and has done anthropological fieldwork in Britain and Ghana. She teaches social anthropology at the University of Sussex, where she also co-directs the prestigious MA in Gender and Development, which is run jointly by the University and the Institute of Development Studies. She is a member of the editorial collective which produces *Feminist Review*.

Kate Young, social anthropologist, was a Fellow at the Institute of Development Studies, University of Sussex for thirteen years and developed the Institute's programme of research and training on gender and development. She is now Executive Director of Womankind (Worldwide), a small development agency which supports women's efforts for development and empowerment in a number of developing countries.

Foreword

It gives me great pleasure to present this practical guide on Gender and Development, for it is a mark of the achievements we are making, within the Commission of the European Community, to translate into practice the Community's policy on women in development.

The EC's policy can be summarized as a determination to integrate women into the mainstream of development on equal terms with men, both as agents and beneficiaries. This policy is based on the principles of human rights and social justice, as well as on the straightforward economic rationale of cost-efficient targeting of resources.

Because women in developing countries play a crucial role in almost every economic and social sphere of life, the integration of women is an issue which relates to virtually every sector of development, ranging from agriculture, animal husbandry, fisheries, forestry, industry and trade, to social sectors like health and family planning, nutrition, drinking water, education, housing, transport and urban planning. Consequently, the Community's policy on Women in Development or the incorporation of gender (to use the modern terminology) affects a wide range of professional staff, both within the European Commission and in the beneficiary countries.

For the EC's policy on gender to be implemented, we recognize that all of these professionals, working within their own fields of expertise, will have to review and adapt their more traditional working methods and goals.

To that end, the Commission has adopted an overall strategy which focuses on increasing the capacity of its development staff to deal effectively with the gender issue. This strategy includes activities like staff training, the preparation of a Women in Development Manual with staff instructions and guidelines; the production of Country Profiles on the role of women in different societies; the engagement of top-level consultants to carry out programming missions to ACP countries which aim to incorporate the gender dimension in community-financed operations, to mention only a few examples.

In this framework, staff training has always been considered a priority and for that reason the Commission decided in 1986 to finance a project with the University of Copenhagen which, in collaboration with the Universities of London and Sussex, prepared a set of training materials on the issue of gender in relation to different sectors of development. These materials were prepared by a number of highly respected scholars in the field of gender studies. No wonder then, that when the materials became available, they were considered of such high quality that we were of the opinion that their distribution and use should not be limited to the Commission services but that they should be made available to a wider public.

Thus it was decided, in consultation and cooperation with the authors, to have the materials edited and published in two different formats.

The first result is a set of seven training modules with accompanying tutor notes for the purpose of 'distance learning', which can be used by interested training institutions and development organizations all over the world. The training modules are being made available through the Institute of Development Studies of the University of Sussex.

The second format is the present book which summarizes the important information, analysis, insights and recommendations of the authors.

I hope this book will serve both professionals and students concerned with development and thereby contribute to a better comprehension of the issue of gender and development, an issue to which the European Community attaches the greatest importance.

Manuel Marin
Vice-President of the European Commission
Brussels, October 1991

Preface

This book is entitled *Gender and Development* because it seeks to explain why using a gender perspective leads to greater success in development efforts. The authors demonstrate how both donors and recipients need to and can incorporate gender awareness into their development-aid policies, analyses, planning and aid administration.

The need to incorporate gender awareness into development efforts was recognized some twenty years ago, when planners began to realize that expecting a country to develop towards modernization with the female half of its population unable to take full part in the process was like asking someone to work with one arm and leg tied behind their back. The first result was the Women in Development movement, which concentrated on identifying the situation of women and aimed to ensure that they had the opportunity to play their full role in the development process.

But the words 'their full role' caused difficulties. What was it traditionally? What is it today? What should it be? Each nation, each region, each culture and subculture has a different answer. Very few elements seem to recur across the cultures. The basic one is the recognition that women bear children and men do not, and a division of rights and responsibilities along sexual lines. Some tasks men will do, others women will do, but which are done by men and which by women vary by culture.

A common element in many cultures is a traditional division of rights and responsibilities which usually places women in an inferior position socially, economically, legally and politically. The two genders share the advantages and disadvantages of their socio-economic groupings, but within these groups, in most cases, women tend to experience further disadvantages relative to men. For example, the daughter of a Brahmin in India will have far better opportunities in life than the son of a middle-class merchant, but in general her opportunities will not be as good as her brother's.

The Gender and Development movement is building on the knowledge and successes of the Women in Development movement. The keystone is the objective analysis of the situation of both genders, recognizing that aside from childbearing, the roles played by men and women are not inherent in biological fact, but are defined by the cultures themselves. Furthermore, many of the functions of men and women in the Third World are substantially different from those in Western industrialized countries.

In seeking to ensure that women play a full role in development efforts, planners and administrators often have difficulty reaching them, learning about their needs, enabling them to take part and ensuring they receive the benefits of development. In fact, according to the authors, due to the lack of gender-aware information and planning, many projects have worked to further disadvantage women in the Third World.

Thus, one of the movement's basic arguments is that in order to be successful, each development effort must be preceded and accompanied by a gender-aware analysis that takes into account the roles and needs of males and females in the area where the project will operate.

As the chapters in this book show, these differences between the genders in roles and opportunities pervade all aspects of life in developing countries. Thus they must be taken into account across the board – in *all* development policies and planning.

The first chapter, on *Gender*, discusses the background and concept of the gender perspective.

To enable women to play a full part in and benefit from development projects, planners and managers must have clear information about where and how women are situated in social, legal and economic terms. Chapter 2 explains how traditional census and information-gathering methods tend to misrepresent the situation and underreport the economic contributions of women in both developing and industrialized countries. It offers new methods and sources for obtaining more reliable information.

Chapters 3 through 6 focus on the gender-related issues affecting women in agriculture, employment generally, housing and transport.

Chapter 7 presents the crucial issue of health.

Chapter 8, on the division of income and management of household resources, weaves together many of the insights of the previous chapters. Household management is relevant to all aspects of development work because reaching and improving the life of the individual is the ultimate goal of every development effort. In this chapter, the reader will see that it is important to start with an understanding of how aid reaches and affects the individual, before looking outward towards the broader aspects of development programmes.

The final chapter provides a cogent analysis of what is required to incorporate the gender perspective into development efforts, and provides practical guidelines on how best to instil gender awareness into the administrations and activities of donor agencies and countries.

The experience of donor agencies over the past twenty years has shown that the application of gender analysis and a gender perspective to development efforts enables planners to understand far better the needs and priorities of both men and women. Through this understanding, aid administrators can improve the impact of their projects and above all help and motivate both genders to work together towards the successful development of their countries.

Chapter 1

Gender

Lise Østergaard

More often than not, development projects sponsored and implemented by Western organizations reflect ethnocentric bias about the sexual division of labour and the family-sharing of income and resources, whether in cash or in kind. The Western middle-class family pattern can be read between the lines: a household composed of a married man and woman and their children, with a male breadwinner as head of the household taking responsibility for supporting the family, and a wife keeping the house, taking care of the children and eventually having a supplementary income from work outside the home. Numerous projects in recipient countries have failed or not fulfilled their goals because planners have ignored the social and family structures in which the development is to take place, and these differ very definitely from the Western model.

Sad evidence today also testifies to the fact that development strategies based solely on macro-economic theories, like structural adjustment programmes have failed to solve the problems of poverty in Third World countries. Moreover they have had the unforeseen side-effects of making the poor and underprivileged even poorer. In Third World countries the 'poorest of the poor' are those rural and urban households headed by women. In addition to grappling with all the other difficulties faced by the destitute, female-headed households have to struggle against the barriers that exclude women from earning a livelihood. These people live below the limit where basic human needs are satisfied – without access to sufficient nutrition, safe drinking water, simple shelter, basic education or primary health care.

This situation has now been recognized by some donor agencies, including the World Bank, and their policies have been broadened to encompass micro-economic aspects of development as well. Development must be a human-centred process, because people are both the ends and the means of development. Therefore programmes of human-resource development must be at the centre of economic development strategy. Everything else – economic growth, fiscal policy, exchange-rate management – is no more than the means to achieve the fundamental objective of improving human welfare. It must also be recognized that women play a crucial role in sustainable development.[1]

Development planning, whether national or international, has traditionally been gender neutral or even gender blind. This was partly because until recently we lacked information about women and their contribution to their regions and because aid organizations were administered with very little insight into gender roles. As a result there was a tendency to marginalize women: development planners have often seen them only as passive beneficiaries of social and health services. Women's active and productive roles in their society were not recognized and not included explicitly in development planning.

Even today the target groups for development projects are often identified as genderless categories, such as 'small farmers' or 'the rural poor'. In the minds of planners these groups are men. In reality many of them are women. It is an implicit assumption that the effects of development projects are potentially beneficial to both men and women. In reality quite often the advantages of development go to the men in the form of increased earnings or labour-saving techniques and the disadvantages go to the women in the form of an increased and unremunerated workload. Should this be recognized, the proposed solution is often the initiation of special 'women's projects', which tend only to marginalize women further as a 'special group' within society. Planners must realize that development goals will only be reached by securing the active involvement of women as well as men, and by bringing women into the mainstream of economic development so that each gender plays its own important role in the process. Specific project components for women – such as credit schemes or training programmes – used as integrated parts of the total plan may be useful, however, as a first step.

WOMEN STUDIES

Research into women's issues is of recent date; studies related to the role of Third World women in the development of their countries have only been carried out regularly during the last twenty years. The pioneering work was Ester Boserup's book, *Women's Role in Economic Development*, published in 1970.

After the appearance of this study, there was a lull and then a remarkable breakthrough in 1978 with a great number of new studies on Third World women. This productivity has continued since then with scientific papers, books, conference reports, and policy documents of various kinds. Methodological developments have also occurred over this period. The early studies were mainly surveys and were later criticized for being descriptive, empirical and non-theoretical. They did, however, serve the important purpose of rendering visible facts about the reality of women's lives, which were formerly unnoticed or invisible, such as the actual economic role played by women whose labour is unpaid and therefore goes unmeasured. Without those early studies, we would not have reached our analytic and theoretical understanding of how women's functions and activities constitute vital parts of the dynamic forces shaping the future of the Third World.

In addition to the quantitative surveys, a tradition for in-depth qualitative studies developed, dealing with women's functions in the local communities, i.e. the household, the village, the urban slum, industries employing women, etc. Although generalizations may be difficult on the basis of such micro-studies, these grass-roots investigations of women's activities unveiled indispensable detailed information about women's roles in production and reproduction. Other studies dealt with global issues, i.e. the influence of international trade and economy on the composition and mobility of female labour.

By pairing small in-depth studies on culture, household, kin relationships and the organization of gender within caste and class with household surveys and urban labour-market studies based on larger samples enabling generalizations about women's actual productivity, we are now able to have a much fuller picture of the organization of gender in social contexts.

Most importantly the combination of quantitative and qualitative studies promoted a concept of development and socio-economic change as a process of transformation, which depends on continuous integration of the cultural, social, political and economic institutions within a given society. This is a view in strong contrast to the unidimensional and technocratic concept of development, which expects inputs of Western technology and know-how into well-defined sectors of Third World societies, to promote economic development and to benefit all automatically.[2]

THE WOMEN IN DEVELOPMENT MOVEMENT

Another type of research has a political as well as a scientific interest in obtaining and documenting factual, empirical information about women on a global basis. They use less formal qualitative research methods, such as participant observation, unstructured and semi-structured interviews and various forms of action-oriented research. These studies do not want to see women only as objects for observation, but to regard them as equal partners who can themselves identify their needs and wishes through the research projects. Thus this type of research not only aims at data collection, but also endeavours to activate and liberate women's own resources and raise their consciousness about their conditions. In action-oriented projects the researchers and the objects for research act as equal partners to identify relevant research topics and to find ways and means of dealing with the problems of women's subordination.

Undoubtedly, the United Nation's International Women's Decade played an important role in this development. The stimulating effect of the three UN Women's Conferences and of the long periods of preparation for them should not be underestimated.

In 1975 the UN inaugurated the Women's Decade at a conference in Mexico City, which helped to spur the 1978 breakthrough of a great number of scientific and popular publications about women and development. As a preparation for the Mid-Decade Conference in Copenhagen in 1980, the UN secretariat took steps to have statistical information about women's living conditions collected through public sources from all regions of the world. It marked the first time attempts were made to provide a global picture of women's situation. The following – frequently quoted – statement about women and the economy originates from that report: 'As a group women have access to much fewer resources than men. They put in two thirds of the total number of working hours, they are registered as constituting one third of the total labour force and receive one tenth of the total remuneration. They own only one percent of the world's material goods and their rights to ownership is often far less than those of men.'

In parallel with the official UN efforts, there were the so-called alternative conferences; the activities surrounding them released an even stronger dynamism. These conferences were organized spontaneously from the grass-roots level and they brought about contacts and cooperation between women researchers and women politicians worldwide. The culmination of these efforts was the Forum, convened in Nairobi in 1985, running simultaneously with the third UN Women's Conference. About 16,000 women met at the Forum to share scientific, political and educational presentations on women's conditions. The dynamism of this grass-roots movement had a strong impact not only on the level of awareness and understanding among people but also on the formation of new research strategies for women studies. Since then research centres have been established in Western as well as in Third World countries and networks have been built through mutual cooperation. In some of the Third World countries, not least in India, some of the scientific groups were at the same time activist groups, fighting not only against the oppression of women in particular, but also against racial discrimination, class dominance and colonialism in general.

SEX AS DESTINY

The Women in Development movement, whether in its scientific, political or popular form, has definitely drawn the world's attention to the fact that women represent powerful human resources in development, that unnoticed they perform the major part of the world's labour and that they do so under very underprivileged conditions.

Women's activities are spread over various sectors of society, productive as well as reproductive. Women's role in biological reproduction and the bearing and nursing of babies is self-evident. It is a false stereotype, however, that because it is 'the biological nature' of women to bear children, it is a natural 'biological' outcome that for their lifetimes women should be obliged to do all the housekeeping and domestic activities.

In all homes the sexual division of labour encompasses both reproductive and productive activities. But the workload connected with the domestic activities which maintain or 'reproduce' daily life are mainly allocated to women, while the more extroverted and distant income-generating activities are allocated to men. This distribution of labour – and of the rights to dispose of the income in cash or kind which results from that labour – is clearly of a social and not of a biological nature.

False stereotypes are pitfalls for thought and lead to irrational actions. It is an ironic consequence of the 'determinism' that accompanies the belief in 'the biological nature of women' that women are often treated as a minority group in spite of the fact that they constitute more than half of the world's population. Phrases like 'this is ɩ for women, youth and other special groups', or 'for women, n and handicapped', have been seen too often in public ents. In this way being female is equated with suffering from an irreversible handicap; women are subordinated or marginalized 'for biological reasons'.

Such biases have not only dominated public opinion. Until fairly recently they have also influenced social scientists to see sex differences as being beyond the scope of social analysis and have thus obstructed their understanding of the social and historical roots of gender relations.

THE CONCEPT OF GENDER

Gender refers to the qualitative and interdependent character of women's and men's position in society. Gender relations are constituted in terms of the relations of power and dominance that structure the life chances of women and men. Thus gender divisions are not fixed biology, but constitute an aspect of the wider social division of labour and this, in turn, is rooted in the conditions of production and reproduction and reinforced by the cultural, religious and ideological systems prevailing in a society.

Among the research centres dealing with women studies, the Institute of Development Studies at the University of Sussex is at the forefront, not least with regard to developing a critical, conceptual and theoretical basis for these studies. Among many other sources, but as one of the first, Ann Whitehead gave a definition of the concept of gender, as a contribution to an IDS conference held in 1978 on the topic: 'The Continuing Subordination of Women in the Development Process' and she based her terminology on collective discussions among colleagues. She outlined the rationale of the concept along the following lines:

> No study of women and development can start from the viewpoint that the problem is *women*, but rather *men* and *women*, and more specifically the *relations* between them.
>
> The relations between men and women are socially constituted and not derived from biology. Therefore the term gender relations should distinguish such social relations between men and women from those characteristics, which can be derived from biological differences. In this connection *sex* is the province of biology, i.e. fixed and unchangeable qualities, while *gender* is the province of social science, i.e. qualities which are shaped through the history of social relations and interactions.

These relations are not necessarily nor obviously harmonious and non-conflicting. On the contrary, the socially constructed relations between the genders may be ones of opposition and conflict. But since such conflicts are not to be analysed as facts of biology and nature but as being socially determined, they may take very different forms under different circumstances.

They often take the form of male dominance and female subordination. The subject matter of analysis is then the various forms that subordination takes, i.e. in women's unpaid domestic work, in the low-wage sectors of the labour-market, in women's work in rural areas in Third World countries, etc. Women have as a whole less control over the family's economic resources; they also have less status relative to that of their husbands in that a greater share of decision-making and authority goes to men.

A gender approach means analysing the forms and the links that gender relations take and the links between them and other wider relations in society. What are the links between gender and economic relations between people, whether members of a household or of different social classes, and how do changing economic relations affect gender relations? What are the links between gender and changes in productive relations and how do the conditions of reproduction of labour affect gender relations?[3]

In short: the concept of gender makes it possible to distinguish the biologically founded, sexual differences between women and men from the culturally determined differences between the roles given to or undertaken by women and men respectively in a given society. The first are unchangeable, like a destiny. The latter are workable and may be changed by political and opinion-shaping influences. The concept of Women in Development is concrete and may lead to marginalizing women as a particular species with inherited handicaps. The concept of Gender in Development is abstract and opens up for the realization of women's productive potentials in development.

GENDER-AWARE PLANNING

The growing awareness of gender issues has now led governments and international organizations granting development aid to set up sections or focal points to be responsible for integrating women's issues into the administration's projects.

Gender awareness should be stressed by all cooperating parties of planning, i.e. in all administrative sectors of the donor administration as well as in the state and local administration of the recipient country. Similarly, gender awareness should be stressed during all phases of the project cycles. It is important to focus on gender issues at a very early phase of the project planning, preferably already in the stage of project identification. Plans of operation should give adequate priority to gender aspects with respect to budget, the recruitment of appropriate staff and the necessity for the organization and training of local women. Not least, attempts should be made to find out from local women what they want and need.

Collection of relevant data about gender issues should be undertaken consistently, and baseline studies be performed in the pre-appraisal phase in order to have systematic knowledge about gender in the project area. Explorative research projects and pre-appraisal missions with broad terms of reference are useful to ensure that projects proposed are truly relevant to local society and culture. In the further course of the project cycle, i.e. operational planning, implementation, monitoring and impact studies, gender dimensions should be investigated consistently and information documents organized so they can be used for the planning of future projects.

In both donor and recipient countries development project staff are usually male and by training oriented to the technical and administrative aspects of the work. This is also true of both expatriate and local personnel in the cooperation countries as well as the appraisal and evaluation teams. A more balanced distribution of female and male project staff and a supplementing of technical and administrative know-how with social, psychological and anthropological expertise would greatly improve an agency's ability to incorporate gender issues appropriately in development planning.

The training of staff is necessary in order to ensure a genuine and appropriate integration of women into development projects. Social scientists require training in how to use their skills effectively and how to work with technical and administrative personnel. The latter need to learn why, how and where it is useful to apply resources from the social sciences.

Gender should similarly be dealt with in relevant sectors of the central administration of donor and recipient countries. In most countries there is a general tendency to push women's issues over into the auspices of the ministry of social services rather than integrating them adequately in relevant ministries like agriculture, planning and finance. Aid agencies should support work on women's issues undertaken by all sectoral ministries and not only that carried out under the auspices of social and health services. This may involve encouraging the placement of officers responsible for women's issues at the same level of competence as other comparable staff in ministries like agriculture, livestock, finance, rural planning, urban planning and cooperative development.

In developing countries women are poorly represented in district development committees and other local planning bodies. Women's roles in their communities are characterized by the separation of male and female spheres and activities and by women's lack of access to the political arenas in which decisions about development are made. Women's needs and wishes are not communicated to the relevant organs of local government and administration. Information about development programmes, opportunities and procedures is not disseminated in ways which make it available to women. Development agencies should actively support adequate representation of women and women's organizations in all committees that deal with development planning and management. Women's chosen representatives should receive support and training so that they can participate actively and equitably in such committees. It is important to remember that, in dealing with bureaucratic procedures, we must do things *with* instead of *for* women, so that they acquire the necessary knowledge and contacts themselves.

In planning project activities, account must be taken not only of the difference between men's and women's situations, but also of differences among women. It is unrealistic to address 'women' in general since age, position in the household, caste and class affect women's ability to participate in development activities. Women's social situation is more strongly determined by family structure than men's. This means that a careful appreciation of variations in family situations is crucial in examining a project's potential and actual impact on women. Women are never simply women; they are daughters, widows, married mothers of small children, unwed mothers, wives of migrant labourers, mothers-in-law. The authority, autonomy, responsibility, obligations and workload they have in the family vary accordingly. And so does their ability to participate in a project and the way they are affected by it. Thus we cannot simply ask how a project has affected women's roles in the family; we must also ask how family roles affect their potential participation in the project.

In particular, development activities which increase women's workload without any proportionate improvement in their situations should be avoided. Any activity that relies heavily on women's labour should include women in planning and managing capacities; wages should be paid for this work where possible and appropriate. Activities that alleviate women's burden of work should be encouraged. Technology that saves labour in unpaid domestic activities, such as food processing and preparation, should be given high priority. Through gender-aware planning, administrators have the chance to mobilize an enormous development potential.[4]

RECOMMENDATIONS

- Planners should be careful about ethnocentric biases, particularly of stereotypes about 'the biological nature' of women and men. Most divisions of labour, of rights and obligations linked to sex are socially determined and therefore changeable.
- Women's work should be reflected realistically in national censuses. In order for this to be possible, micro-studies in the form of spot tests should be performed and compared with macro-studies.
- It is crucial to deal directly with the basic living conditions in the regions of development; this type of information is often provided by a combination of quantitative and qualitative studies reflecting development as a process of transformation which implies cultural and social as well as economic aspects.
- Women's potential for productivity is part of the human resource base in any society and should be dealt with by bringing women into the mainstream of economic development.
- Planners should realize that development is a multifaceted process which cannot be pursued by unidimensional and technocratic measures alone. Development is a human centred process since people are both the means and the ends of development.

NOTES

1 World Bank (1989).
2 Bleie and Lund (1985).
3 Whitehead, in Young *et al.* (1979).
4 Whyte et al. (1987).

Chapter 2

Statistics

Alison Evans

To ensure that women are adequately represented in development policies and programmes, and to promote greater equality for women, it is essential that we know where women are placed in social and economic terms. How else can we plan or measure progress? It is important to know where different categories of women are placed *vis-à-vis* men and also *vis-à-vis* women in other socio-economic groups, for women are no more homogeneous than men.

Yet it is generally agreed that the statistics available for most Third World countries fail to provide this information. Until the 1970s mainstream research, policy and planning virtually ignored the economic role of women. The dominant view was that women's participation in society was outside the economic mainstream and mainly restricted to activities for which they were stereotypically most suited: family and child welfare and household work. The conceptual categories and techniques used in data collection were conformed to this dominant view, so the statistics collected reflected these stereotypes.

Qualitative research and case studies carried out in the 1970s began to reveal the crucial economic role taken by most of the women in the Third World. Today, most experts in this field would agree that much of the data available on women in the Third World is not only inadequate but also presents a distorted picture of women's contribution to development. With the inauguration of the UN Decade for Women in 1975, priority was given to disaggregating by sex all national economic and social statistics to make visible the full extent of women's participation in economic and social life and their true status in terms of income, health and education. Progress towards achieving this aim has been slow and uneven either because cost implications are too great for some countries or because there is a lack of commitment on the part of national statistical offices to implement such changes within an acceptable time period.

Disaggregating data by sex is a necessary requirement for improving the data base on women. However, it is not a panacea for improving the visibility of women; it only uncovers the tip of the iceberg. The implicit assumption underlying the call for greater disaggregation is that conventional conceptual categories and techniques for data collection hold the same meaning for everyone.

As this chapter attempts to show, this assumption is by no means always justified. Censuses and most census-type surveys are based on systems of data collection that were originally developed for industrialized market economies. These systems contain assumptions about the functioning of the market economy and the value of economic activity that exclude the contributions made by the informal economy and undervalue economic activities carried out beyond the market place. This has the effect of undercounting and undervaluing the contributions of both women and men in Third World and industrialized countries, but the effect is altogether greater for women in the Third World (see below).

It is also assumed that the methods and techniques used to elicit data work equally well for men and women. Yet evidence shows that men's and women's experience of their economic and social environments differs and this affects how they respond to questions about their situation. For example, in some cultures women will respond quite differently to questions about their participation in economic activities depending on whether their husband or father is present or not.

Thus even though data on women *are* increasingly available at national and local levels for researchers and planners to see, the statistics do not necessarily give a true picture of the areas they claim to measure. It is important that both the users and the producers of statistics recognize these problems and try to overcome them.

MEASUREMENT TECHNIQUES

To be described statistically, an activity or characteristic must be susceptible to quantitative measurement. This is neither an easy nor necessarily a neutral procedure. The measurement techniques chosen, as well as the dominant perceptions and attitudes about what information is important or relevant, exert a strong influence on what is actually measured by statisticians and what is not.

For example, census and national income accounting methodologies have been severely criticized for excluding production that is not traded on formal markets.[1] This is particularly significant for the Third World countries where both the volume and value of production for own-consumption and for informal exchange is still substantial. Estimates of the national product that exclude such production underestimate the full value of national economic activity. The problem is most acute when it comes to recording women's activities. As Boserup puts it, 'the subsistence activities usually omitted in the statistics of production and incomes are largely women's work'.[2] An example can be drawn from an analysis of the 1976 Peruvian Peasant Survey which shows that once 'non-market' production is measured the proportion of women identified as being engaged in agricultural production rises to 86 per cent compared to the original figure of 38 per cent.[3]

Other factors that influence the accuracy and coverage of both quantitative and qualitative data collection are the timing of interviews, the length of the reference period and the language in which the interviews are conducted. Choice of an unseasonal time of year and a limited reference period are among a number of technical explanations for the underreporting and misrepresentation of women's work in agricultural statistics. For example, census enumerators usually visit rural communities in the dry season, when access is easiest. However, most rural communities are busiest during the rainy season. If the standard one- or two-week reference period is used, this work will be excluded from the record. If rainy season activities are probed, the accuracy of the answers will depend upon the recall and the perceptions of the person chosen to respond. There is some debate as to how well people recall annual patterns of work; both under and overreporting of work is observed. Research indicates that underrecording of work is particularly acute for women whose current activity at the time of enumeration (a single reference period) is not on-farm work but household/domestic work. As a result, they are classified as 'non-workers', even though their main activity is on-farm work at other times in the agricultural cycle.

Conceptual categories

One important reason why present statistics, even when disaggregated, can give a distorted social and economic picture is the choice of conceptual categories used in data collection. For example, the standard concept of economic activity refers to participation in a *productive* activity. Economists and planners rarely consider work associated with the household and own-consumption to be productive activities. This directly affects the recording of many aspects of women's work.

A study from India shows that if the 32nd round of the National Sample Survey had classified all the activities assumed to be 'domestic' as 'economic' – for which there is a valid case given the high economic content of many 'unpaid' domestic activities – the labour-force participation rate for rural women above the age of five would have risen from 30.5 per cent to 52.3 per cent, compared to a male rate of 63.7 per cent.[4] Information on time spent in unpaid work is crucial for development planners to know if they are to succeed with projects and programmes that make specific demands on 'available labour time'. Yet despite such arguments, there is a general reluctance to go beyond simply disaggregating data and to make basic changes in statistical conventions to meet new arguments about recording women's unpaid work as economic activity.

This reluctance reflects another conceptual obstacle: *the ideology of sex-difference*. There is a widely held view that the sex of a human being brings with it particular attributes that predispose him or her to a particular 'role' in society: in reproduction, production, kinship, marriage and the family. Women are assumed to be 'naturally' suited to caring and reproduction, while men are 'naturally' suited to being providers and protectors. The sexual division of labour is therefore based on the biology of human reproduction and forces men's and women's activities to be guided by 'sex roles'.

This way of seeing the division of responsibilities between men and women gives rise to a number of problems. One is the assumption that the divisions between men and women are somehow pre-social and therefore fixed; that the sexual division of labour is essentially the same for all human societies at all moments in time. A scant look at history proves this is not so. The division of labour is not immutable: it has the capacity to change and adapt to differing social and economic conditions. Biologically based explanations of the division of labour and the unequal positions of women and men in society fail to provide adequate explanation of the wide and varied contributions that women make to society; nor can they provide an adequate basis for action to promote greater equality for women in the process of development.

Moving towards equity – the gender model

To obtain more accurate data on women and an objective picture of what is actually happening in a country we have to use a model of human society that goes beyond sex roles. The equity-oriented or gender model views the life of women as being conditioned by *non-biological* factors, such as political and religious ideology, culture and the economic system. Categories of men and women are recognized to be heterogeneous and are further differentiated by social divisions such as class, race and ethnic group. According to the gender model, quantitative and qualitative information about men and women should be organized so as to reveal basic differences both within and between groups of women and men. This may be a matter of stratifying labour-force data not only by gender but also by socio-economic group, or cross-tabulating data on the educational performance of women by race and ethnic group. Such procedures immediately enhance the meaning of the data and make possible a more meaningful analysis of the socio-economic position of women. Not only is it important to enhance the meaning of data through the reorganization of data sets, it is also important that statistical information about women and men is analysed in relation to its social, economic and political context. Again the gender model is explicit in its criticism of statistical information taken in 'isolation' and emphasizes the importance of comparability between data sets in order to build a more holistic picture of economic and social phenomena. For example, comparable data on women's participation in productive activities have limited meaning without comparable data on levels of income, education and participation in 'reproductive' activities, such as domestic work and child care.

General arguments about technical and conceptual issues in data collection are mainly aimed at the producers and users of censuses and national surveys. However, there is a wide range of different data sources and each kind presents its own problems. Despite the efforts of the various UN agencies to standardize definitions and methods across the globe, data continue to be collected at many different levels, in many different ways and for many different purposes. Thus concepts and methods vary across countries, within countries and between surveys themselves. Table 2.1 (next page) gives some idea of the different types and sources of data that can be found in Third World countries. Variations exist in the depth of data coverage across countries and in many instances only a limited selection of different kinds of data are actually available or kept up to date.

Faced with arrays of statistics which measure different things, which may be inaccurate or out-of-date, which may or may not be disaggregated, and which were probably collected using sex-biased conceptual categories, how can we try to gain an accurate picture of

Table 2.1 Summary of some major sources and types of data

Level of collection or presentation	*Published sources*	*Types of data available*
1 International statistics	International statistical yearbooks, produced by the UN statistical office and by various UN agencies e.g. FAO, ILO, WHO, UNESCO. World Development Reports and International Financial Statistics produced by international development institutions, e.g. World Bank, IMF. Data are drawn mainly from national sources but presented to allow comparison of national development indicators.	Mostly annual data covering population, fertility, mortality, health, education, economic activity, labour force and employment, industrial and agricultural production, financial information, etc., and broad development indicators, such as quality of life indices, 'status' of women, etc.
2 National statistics	National censuses: produced by the Central Statistical Office (CSO) or a Government Statistical Department. Data are gathered, analysed and published according to an internationally agreed format under the auspices of these departments.	Mainly decennial data covering urban and rural areas on population, economic activity, labour-force participation, housing, education, etc.
	Sectoral censuses and surveys: produced by CSO from small-sample surveys.	Sometimes annual, often 5-yearly to supplement decennial data. Covering agricultural and industrial production, employment, labour force, etc.
	Household surveys produced by CSO from sample surveys, often in collaboration with external agencies e.g. USAID/UN agencies.	Rarely annual, usually intermittent or one-off surveys covering household composition, income, expenditure, consumption, education, health, economic activity – usually urban *or* rural (but not both).

Table 2.1 Summary of some major sources and types of data (continued)

Level of collection or presentation	*Published sources*	*Types of data available*
	Administrative data and civil registers: produced by National Records Office often drawn from local data sources.	Continuous series of vital statistics, such as births, deaths, marriages, etc.
	Archive data: collected by National Archives Office drawn from government and non-government sources.	Historical data on wide variety of topics, including both quantitative and qualitative information, e.g. ethnographics.
3 Local statistics	Small-scale surveys produced by local government departments. Surveys run by projects in the form of feasibility studies or monitoring exercises. Local data collection by clinics, hospitals, schools, churches.	May be collected continuously or erratically as specific information is required. This tends to make data series incompatible, making comparative analysis difficult. Data usually found on agricultural and food production, employment and time allocation, health and vital statistics, education and sometimes on specific topics, e.g. women in food production, women in rural employment, etc.

Note: Some sources may be unpublished.

women's situation in a country? The rest of this chapter examines the problems in more depth and proposes some solutions.

FINDING INFORMATION FROM THE DATA

In this section we look at the way conventional sources provide statistics on households and families, education, economic activity and income. In each area we outline the main gender issues involved and suggest ways in which the data might be modified to take better account of equity considerations.

Households and families

In censuses and surveys, the household or family unit is central to the collection of data on economic activity, income and expenditure, as well as social questions such as education, fertility and migration. The definition of the household unit is therefore crucial. Many social scientists argue that the standard household definition, based on a group of people living under one roof and sharing the same cooking pot, conceals more than it reveals about the composition of household groups and the welfare of household members. The effects of this standard household definition on the quality of household data vary from country to country and even from survey to survey. However, there is now wide recognition that the standard conceptualization of the 'household' masks vital detail on the diversity of household types and the characteristics of women's work and women's socio-economic position within households.

The household as a unit

Under international guidelines, individuals who are all sharing a common kitchen or the same cooking pot and living under the same roof are usually considered to constitute one household (what we call the 'standard definition' of the household). Most censuses try to document different forms of household types, so some assumptions are clearly necessary. Yet many of the problems that arise in enumerating households result from the lack of fit between the simplified categories of household type available in censuses and national surveys and the diverse household characteristics which in fact exist.

Some of the assumptions currently in use are that:

- Households are constituted around family-based relationships centred on marriage and parenthood
- Co-residence is a defining feature of the household
- The residential unit and consumption unit are one and the same
- The household is the unit that pools and distributes resources.

In practice households are not necessarily family-based. It is not uncommon for households to be organized around production and/or consumption arrangements which can include any number of non-family and visiting members. For example, in some rural areas the residential unit and the productive unit are also the site for family consumption but in other areas family members are spread (either on the basis of kinship or age and gender) across a number of 'households' for consumption purposes. Consequently there is no necessary correspondence between family group, residential unit and consumption arrangements.

A study of working-class households in Bombay provides an excellent example of the complexities involved in applying any standard concept of the household to even a small sample of households in a non-standard culture:[5]

- Many people in the sample ate together regularly but did not sleep under the same roof, so it was not possible to use the *residential unit* as the basic concept.
- In other cases, people shared the same roof but co-residents did not pool individual incomes nor share their resources. This disqualified the definition of the household as *the unit that pools and distributes resources*.
- A spatial separation between the town and countryside split other groups, yet the respondents still considered themselves as single family units. In these cases income and consumer goods were not distributed from a single location by a single household member; instead, resources and family members were constantly shifting between separate locations.

Thorner and Ranadive also had difficulty in determining just how many adults and children should be counted in each of the sample residential units because the number residing in each house tended to vary by season, changing economic circumstances and the life cycle of the family.

In reality, households and family groups take a variety of forms, depending on cultural and material factors. There is no single answer to the question of how to define a household or the relationships within it. The decision should depend on whether the household is viewed primarily as a kinship unit, a residential unit or an economic unit, and which definition will provide the most useful information for the purpose at hand.

Changes in household composition

One of the great shortcomings of census data is the lack of information on the relationship between changing socio-economic conditions and changes in household structure and composition. This information would greatly enhance our understanding of the relative position of women and men within households and how changes in household composition relate to changes in women's unpaid and paid work, their education, health and fertility.

To provide the information needed for such analyses, data not only has to be disaggregated by sex, but also by the following categories:

- *Organization of household authority*: joint, female or male-headed households, with or without spouse/partner present
- *Type of marital union*: legal and consensual, polygamous, monogamous
- *Household composition*: nuclear, joint, extended, two or three generational or other.

Small-sample and multi-purpose household surveys are useful for this purpose because they can codify the data by household type after it has been collected. For example, in her study of urban Bengali households Standing identified at least ten main categories of household types and several sub-categories.[6] She was then able to relate household relationships to information on marital status, employment, income, consumption, fertility and changing household composition over time. While undoubtedly valuable, this process of collecting data is inevitably more time-consuming and makes coding and tabulating data more difficult.

Family relationships: the concept of 'head of household'

Designating a person as 'head of household' is a short-hand way of identifying family relationships, since all others can be described in terms of their relationship to the head. In theory, household head is simply a reference person. In practice, the question of who is selected as head of household is strongly biased in favour of the male. Moreover, the designation of a person as 'head of household' can strongly influence the gathering of information about the rest of the household.

Selecting the household head

The choice of criteria for selecting a household head should be influenced by the purposes of a survey. A survey primarily aimed at gleaning economic information, for example, may benefit from using different criteria than a survey designed to gather a particular kind of social information. In practice, however, respondents and census enumerators (who are mostly men) decide who is nominated head of household. As culture and ideology in almost all societies tend to privilege male status, headship is almost always attributed to a man regardless of a woman's economic or social role.

This fact is illustrated in Southern Africa where adult men migrate from the rural areas for long periods in search of wage work on commercial farms, in mines or in cities. Women generally remain behind and have the responsibility for managing the rural household, maintaining the farm and providing the daily food needs of the family. It is women, therefore, who are effective or *de facto* heads of households. Nonetheless, under conventional data-collection methods, the male migrant or some male relative – by virtue of their kin status and gender – would be recorded as the nominal and *de jure* head of household. This not only gives a false impression of the organization of the household, it also obscures the responsibilities and burdens the woman is bearing.

Headship bias has made it difficult to determine the number and composition of households headed by women, whether *de facto* or *de jure*. It has also masked the circumstances under which women-headed households are emerging in rural and urban sectors. For instance, are women-headed households a reflection of increasing male migration, or the fragmentation of households due to economic hardship? This information is extremely important to obtain as these households are often among the poorest households in both rural and urban areas.

Another serious practical implication of headship bias is that it obscures the crucial role that women play in controlling and allocating household economic resources, and consequently the importance of targeting development aid at women as well as men. In policy terms, a woman who is bearing the household's major farming responsibilities could find that she is not eligible for participation in a training scheme for local farmers because all the farmers targeted for assistance are household heads and assumed to be men. Even if her husband attends the programme, there is no guarantee that he will pass the knowledge on to her, that she will then be able to put it into practice on a daily basis or that it will be relevant to her own farming needs.

Knowledge about female-headed households is on the increase, but these data have mainly been collected from small-sample household surveys and special studies – not from censuses. There is a real need to have headship data disaggregated by sex, age and marital status. The UN department INSTRAW, which is concerned with research on women, has drawn up technical specifications for improving the database on women and has suggested that surveys and possibly censuses should distinguish female-headed households in the following way:

- *Female headship by household composition*: that is, due to the absence of an adult male through migration, divorce, separation or death
- *Female headship by economic responsibility*: that is, households in which a woman is the sole or main economic provider regardless of whether a man is or is not present.[7]

Another important aspect of the headship problem is the tendency to identify all household members with the characteristics of the household head. Thus the economic activity of the male household head is used to identify the activity status of the household as a whole, irrespective of what the other members may be doing or the income they are contributing. In addition, the well-being of the household head is invariably treated as an indicator of the welfare of all household members. It is assumed that all household members share a common standard of living with the household head and that there are no inequalities in well-being between household members. Minimal information is consequently available on differences in income, consumption or educational levels between women and men within the same household.

Implications for planners

Despite these major problems, the concept of a reference person or household head continues to be a necessary tool. It is central not only to the collection of data but to the targeting of development assistance and certain kinds of welfare payments.

There are several steps that planners can take to overcome these problems:

- *First*, they can supplement conventional household data with information on female-headed households and the composition of household types.
- *Second*, they can set the criteria for choosing the reference person in censuses so that it is easy for enumerators to follow and avoids the possibility of automatically choosing the adult male as the household head.
- *Third*, wherever possible, data from all adult members of a household or family should be collected and presented. Researchers can bypass

the problem of designating a household head by using multi-purpose surveys and sample surveys to collect detailed information on the separate characteristics of household members.

- *Finally*, researchers can use sample surveys to distinguish with more precision the characteristics of poor households in terms of household size, gender composition, educational levels and income status of all members.

EDUCATION AND TRAINING

Data on educational attainment and literacy in the Third World are some of the most widely used statistics in development work. Planners and policy-makers seek information on the level of human resource development – which is intimately linked to the level of literacy, knowledge and skill in a society – to determine the potential for economic growth and development. Researchers often use data on educational attainment as a proxy for class position. For many people, education also indicates socio-economic well-being.

A growing literature documents the incidence of gender discrimination in education. This operates at many levels. In the household, cultural norms and expectations often operate to the disadvantage of girls who want an education, especially beyond the elementary level. External pressures, like the introduction of school fees, have resulted in families opting to keep sons rather than daughters in school. In the schools themselves girls are often offered a different and inferior curriculum and are subjected to bias in selection procedures as well as unconscious discrimination from teachers.

For developing countries to make full use of their human resources, the bias and the backlog of discrimination must be overcome. To accomplish this planners and policy-makers need reliable information on the extent of discrimination, which groups are currently most affected and which groups are most vulnerable for the future. The data currently available in censuses and education records can provide some of this information, but there are a number of problems.

Data problems

Population censuses provide useful benchmark data on education attainment and enrolment. However, this information refers only to the *stock of education* in a country – literacy levels, total number of school years completed, the proportion of the population with certificates of various kinds. Very little information is available on the *flow of education* – attendance, transition and retention rates, continuation data, specialization, non-formal education.[8]

International data sets compiled from national records by UNESCO and the UN Statistical Office offer international comparisons mainly of stock variables. This means that comparisons can be made of levels of educational enrolment disaggregated by sex and age cohort, but less systematically on school attendance. A further restriction here is the differences between nations in the concepts and methodology of data collection. For example, enrolment and attendance data tend to vary depending on whether it is compiled from census tabulations, household survey data, or school/administrative records. Enrolment data from two countries cannot be compared unless standard definitions are applied at the national level to ensure that the level and length of enrollment are strictly comparable.

Another problem is that insufficient data are available on educational performance over the life-cycle, cross-tabulated by sex, age and socio-economic group. The concepts and methods of data collection are generally not sufficient for assessing links between education, employment, occupational mobility and income status, and how these relationships differ for women and men.[9] Both national census and international statistics concentrate almost entirely on the official school system. Thus statistics on vocational training, non-formal education and functional literacy are not easily obtainable.

Suggestions for improving the data

A UN/INSTRAW document in 1984 earmarked a number of key areas in which data on education and training could be improved. However, improvement requires disaggregation by sex, and the changes that have to be made to accommodate this are probably beyond the scope of censuses and national surveys. Instead, it is probably more feasible to extend current education records and multi-purpose household surveys to gather more information and to conduct qualitative studies of educational performance. Such types of data collection are not yet available in all countries. Even where available, they are often not coordinated with each other, making comparisons between different sources difficult. None the less, every effort should be made to enhance these supplementary data sources and coordinate them with available demographic and socio-economic data.

The main areas of information where it is most important to build up disaggregated statistics are access to education, continuation, the provision of educational services, attainment, non-formal education, the teaching profession and the relevance of education to career options.

Access to education is normally measured through data on enrolment, reflecting the proportion of eligible age cohorts who are registered in a particular educational level. Two problems with this measure are that it is usually expressed in absolute terms rather than relative terms for women and men and that it takes no account of corresponding attendance rates. Educators and planners need disaggregated school attendance data to match up with annual enrolment data. With it they can monitor the opening and closing of educational gaps in each age/gender cohort. They can also evaluate the effects of differences in access to education by location and socio-economic group.

Data on *continuation* give some indications of the capacity of certain population groups to remain within the educational system. *Completion rates* record the numbers of boys and girls who complete a stage in the educational cycle. *Retention rates* record the number of students who remain within the system, thus indicating the rate of school drop-out and the stages at which drop-out is most or least likely. *Transition rates* record the numbers of girls and boys each year who make the transition between primary, secondary and higher education. Planners need such data in order to study and assess the differential effects of various policies, like the encouragement of equal educational opportunities or the introduction of school fees.[10]

Information on *educational attainment* comes mainly from completion data: the number of girls and boys who have completed specific levels of education and the absolute numbers of school years for which each woman and man was enrolled (regardless of achievement). Planners use attainment data to examine different forms of inequality in an educational system. Data on years attained can be compared cross-nationally, but the comparison of levels attained depends on the way these are specified in each country. UNESCO has tried to overcome this problem by specifying six standard levels of education applicable in most countries.[11]

The provision of educational services is another way planners and policy-makers can estimate the extent of equal educational opportunity for women and men. Typical indicators here would be the numbers of sex-segregated schools and the prevalence of sex-differentiated programmes and curricula within schools. Such data can usefully be integrated into annual household and educational surveys to help to measure the size of the gaps in provision of schools and courses for girls and boys. However, these quantitative measurements are much more useful when they are analysed alongside other information, such as qualitative information about the content of courses and the performance of the genders in sex-segregated versus co-educational schools.[12]

Disaggregated data on the teaching profession – the training and size of the stock of women and men teachers and the subjects they teach – can help to determine whether there is equality of opportunity in the educational system and whether it is linked to the performance of girls compared with boys.

Within *non-formal education services*, data should be gathered on attendance and completion rates and the availability of options by age and gender, to allow a judgement on whether literacy, adult education and vocational programmes offer equal opportunities to women and men.

ECONOMIC ACTIVITY AND LABOUR-FORCE PARTICIPATION

Global estimates in the 1980s suggested that the total female labour force had doubled since 1950 with women constituting one-quarter of the world's industrial labour force and around two-fifths of its agricultural labour force. This trend can be explained by global processes of socio-economic development that are leading to changes in economic structure and social organization. These in turn are generating a larger demand for and supply of women's labour. In addition, decreasing fertility in some regions, reduced mortality and increasing rural-urban migration are both pushing and pulling more women into economic activities outside the household sphere.

In order to respond to the labour-force potential of women, policy-makers need to know what work women are doing now in market and household spheres, how women's work in the market interacts with their work outside it and what factors are influencing women's entrance into formal economic activity. To obtain this information policy-makers and their advisers usually turn to the population census or to census-type surveys on labour force and employment. But as we mentioned above, there is a great deal of controversy surrounding the measurement of economic activity from these sources. Censuses in particular are considered a poor guide to economic participation beyond the market place and especially to the diverse economic participation of women.

Most censuses today use the labour-force approach to economic activity and define the economically active population as 'all persons of either sex who furnish the supply of labour for the production of economic goods and services' during a given reference period.[13] In this definition *economic goods and services* means those *presumed to contribute to economic growth*. Active labour is thus measured in terms of its links with market activity.[14]

However, the activity of a large proportion of the population in developing countries is only loosely or informally linked to the market. Thus a significant amount of productive work is not enumerated by national accounting methodologies and estimates of the active labour force tend to exclude workers who are not formally engaged in market-oriented activities. The underenumeration is particularly relevant to estimates of women's work in terms of both its extent and value and often leads to serious underestimates in censuses of the number of women in the official labour force.

Employment status

The definition of employment status is important because it underlies the information available to policy-makers on movements in the structure of employment and trends in the status of the workforce. The tightness of a census definition of economic activity will determine the proportion of the population classified as self-employed or unpaid family labour. The nature of this definition varies by country. For example, the censuses of Malawi, Nepal and Tunisia classify almost all economically active farm women as self-employed, whereas the censuses of Gabon and Thailand classify them as unpaid family helpers. Earnings are conventionally used to mark the borderline between self-employment and unpaid family labour. But in many cases the minimum level of earnings needed for self-employed status is not made clear. Nor is it clear whether a woman who trades family produce on an irregular basis is self-employed or not.

Recently researchers have laid emphasis on the recorded shifts in the profile of female labour-force participation. For example, parts of rural India are marked by a widespread shift in female employment from cultivator/wage labourer to unpaid family labour. This shift has been linked to the proliferation of green revolution technologies and a switch to cash-crop agriculture. However, in other countries a reverse shift is apparent: women are moving from unpaid family labour to agricultural wage labour. Planners and policy-makers must ask whether and to what extent these shifts are real or reflect changing *definitions* of employment status as suspected by many researchers.[15] We should also ask whether the changes reflect shifts between part-time and full-time work, temporary and permanent employment and whether the shifts of females are complementary to or in conflict with trends in male employment status.

Unemployment

Unemployment figures usually refer to 'open unemployment', which is defined as all persons available for and 'actively seeking' paid work (over a given period). This is really an ill-fitting concept in most developing economies and generally leads to underreporting of unemployment. Some countries take a strict line on the definition of unemployment: they rigidly apply the 'active search' clause and operate a short reference period of usually one week. In such circumstances potential workers who are available for paid labour but are without the means or encouragement to search for it are likely to be classified as 'inactive' labour rather than active and therefore unemployed. This is frequently true of women who, although technically available for paid work, are discouraged from actively seeking it because of social and cultural factors such as lack of training and education, family resistance or a perceived lack of opportunities. Labour-force studies indicate that if the search criterion were relaxed many discouraged workers would be classified as unemployed. If this kind of 'concealed' or 'disguised' unemployment were counted in labour-force statistics, it would have a huge effect on unemployment figures and would possibly put greater pressure on policy-makers to implement job-creating programmes targeted at unemployed women as well as men.

Underemployment

The concept of *un*employment refers only to those workers unable to find paid work. The criterion for *under*employment is either:

- working fewer hours than the number considered to be 'normal', or
- Working for 'normal' hours but for disproportionately low earnings.

Few countries have been able to include measures of underemployment in their censuses, partly because it remains something of an elusive concept. One study in Colombia indicated that underemployment was significantly higher for women than for men in the middle age ranges.[16] In Java, rates of open *un*employment were greater for men, but rates of *under*employment were much higher for women.[17] Given the state of existing statistics and the problems with applying the concept of *under*employment, it is not clear whether these differences between women and men are real or caused by attitudinal bias in the evaluation of women's and men's work: that is, what may be considered *under*employment for a woman may qualify a man as being *un*employed.

The concept of *under*employment is particularly difficult to apply to women who are responsible for heavy domestic work. What is a 'normal' working day for women who are involved simultaneously in housework, childcare and unpaid productive work? Are they overemployed because of the hours involved, underemployed because of their disproportionately low earnings, or unemployed because given the chance they would be available for paid work? These kinds of questions remain unresolved in current statistical methodologies.

Methodological discrepancies in data

Although censuses and census-type surveys generally adopt UN recommendations on the definition of economic activity, discrepancies among data sources can have profound effects on both the absolute and relative calculation of labour-force participation. For example, in Egypt, the 1960 population census reported that women represented only 4 per cent of the agricultural labour force, while a detailed rural labour-force survey recorded women doing 25 per cent of all productive work in farm households. Similarly, a change in the definition of economic activity in India's 1961 and 1971 population censuses reduced women's labour-force participation rate by 23 per cent.

Almost without exception, compared to surveys of labour force and employment, national population censuses count fewer people as 'economically active' and disproportionately fewer women as economically active than men. This is highlighted in Table 2.2, which compares various sources of data on the agricultural labour forces of five countries.

Improving the information about women

It is frequently argued that one of the first steps needed to improve the measurement of women's economic participation is to expand the definition of economic activity so that it includes work that presently lies outside the 'production boundary'. Women spend a great deal of time in expenditure-replacing activities which involve the production of goods and services, mainly for consumption by household members, that would otherwise have to be bought in the market.

Table 2.2 Comparison of various sources of data on the agricultural labour force in selected countries

Country	*Source*	*Year*	*Ag. labour force (Activity rate)*	
			Female	*Male*
Cameroon	Pop. census	1976	41	48
	ILO estimates	1970	55	62
	FAO ag. census	1972/3	63	50
Malawi	Pop. census	1977	57	54
	ILO estimates	1970	47	74
	FAO ag. census	1968/9	86	83
Iraq	Pop. census	1977	10	16
	ILO estimates	1970	1	35
	FAO ag. census	1971	27	39
Sri Lanka	Pop. census	1963	12	31
	ILO estimates	1960	16	38
	FAO ag. census	1960	37	49
Brazil	Pop. census	1970	4	36
	ILO estimates	1970	4	38
	FAO ag. census	1970	17	36

Source: Dixon-Mueller (1985, p. 94.)

The discrepancies in the statistics in Table 2.2 result from variability in the precision with which economic activity and the 'production boundary' are defined; the length of the reference period (anything from one day to two weeks); the minimum age for the group considered (7, 10, up to 15 years); and the minimum number of hours of work to qualify as active labour (3 hours per day or 10 to 20 hours per week).

Using data from the Philippines, King and Evenson give a

convincing account of the implications of counting 'work' as both market work and home work.[18] For example, they found that home production (expenditure-replacing activity) and market production account for equal shares of Filipino households' total real income. Husbands contribute most in terms of average market income but women and children between them contribute approximately 50 per cent of this income from their participation in unpaid family labour.

However, a number of difficulties arise when trying to measure household production for own-consumption. First, a decision must be made as to what should or should not be counted. This amounts to choosing between counting the economic value of (a) all work done within the household, including caring for children, or (b) counting only restricted economic services that correspond or contribute to activities in the market, such as food processing, gathering firewood and water, craft activities and so on. Second, because household work tends to merge imperceptibly with the process of living, to redefine the concept of productive activity we must have systematic observation of the time spent on household activities.

Time use and time allocation surveys offer one way of collecting more detailed and more accurate data on women's work. They can be used to measure the flow of labour between productive work, household chores and leisure on a daily, weekly or even seasonal basis. In particular, time-use surveys:

- Identify primary, secondary and tertiary occupations
- Include productive activities on the borderline between economic and non-economic categories, and
- Permit analysis of gender and age specialization and trade-offs within the household unit.[19]

There are a number of methodological difficulties involved in conducting time-use surveys, but as research methods develop, the techniques for collecting such data improve. National surveys may soon be able to include questions on time use, but with less detail than intensive studies. To measure trends and variations in economic activity, unemployment and underemployment (especially in agricultural areas), it may well be sufficient to have one-off interviews that estimate the approximate time spent in a limited number of activity categories.[20] Aggregate information can then be supplemented with more detailed regional or community studies of economic activity and time allocation.

Anker suggests another way of improving the quality of data on economic activity.[21] He proposes using four categories of economic participation:

1 *The paid labour force*: employees for cash or kind
2 *The market-oriented labour force*: the self-employed (unpaid family workers, employers, own-account workers, members of producer cooperatives) who produce for the market, with no minimum hours specified
3 *The 'new standard' labour force*: those involved in household-oriented productive activities, such as tending livestock, processing crops for home production and preparation of meals for hired labour
4 *The extended labour force*: those involved in gathering fuel or water and making household substitutes for purchased goods and inputs. (Note that housework and child care are *not* included.)

Anker argues that a simplified time-use survey can provide the information needed to make these labour-force distinctions. These categories are particularly useful because they distinguish between different *levels* of economic activity. In practice, respondents will fit into more than one category. Time estimates allow for this, as well as providing information for prioritizing each activity in terms of primary, secondary or tertiary importance. A variety of labour-force measures can be estimated for each category, such as basic economic activity rates, age-specific rates and the gender ratio of workers. Collected over time, the aggregate data can reveal how women's work is changing compared to men's. For policy purposes, Anker's categories allow valuable comparisons of the size of the self-employed labour force engaged in home production and the paid labour force, and how they differ by gender, age and other factors.

INCOME

Ferreting out information on women's incomes from developing countries' statistics on national and per capita income is a task that has hardly begun. There are a myriad of difficulties to be overcome:

- Most conventional statistics focus on personal and household income derived from employment, self-employment, property and transfers from the state or other individuals. Very little of the national data in developing countries has been disaggregated by sex, so it is very hard to measure women's income separately from men's or the total household.[22]

- In permanent, formal sector occupations, rates of pay and hours are identifiable and recallable. But for many workers in urban and rural areas, wages are not the primary source of income. Even when wages are paid, they often come intermittently, not always in cash, and at rates that vary by season and gender. This irregular, heterogeneous income is hard to measure.
- Information on women's incomes is often not recorded because of a general lack of recognition of women's economic activity, a bias in favour of monetary income and the tendency for national statistics to lump the contributions and claims of women and children into catch-all categories such as 'household income' or 'total farm income'.

Most of what we do know about the nature and extent of income generation by women – both in cash and in kind – comes from case-study material and very specific qualitative material. The reality revealed by this micro-economic research is far more complex than the census statistics allow for with their simple categories and assumptions. For instance, most urban and rural households have two or more income earners: the husband and wife(s), the father and child(ren), or the mother and child(ren). Many households do not pool all or even some of their income. Some households keep incomes separate, but contribute towards joint expenditures. Others split incomes between two or more households: e.g. between the marital and natal home, or across the households of co-wives. In poor households the split between cash and kind income, such as food, is shown to be critical to household survival strategies.[23]

In general the case studies show that in most developing societies women make significant income contributions to households from their activities in the home and in the market. This income is not necessarily in cash, nor is it necessarily controlled by the woman who has earned it. When women do retain control their income does not always go into a common fund, nor is it necessarily declared accurately. However, research on poverty groups has shown that women allocate a greater amount of their income than men to the basic survival needs of their children.

From a policy viewpoint, then, realistic efforts to raise the general standard of living of women have to treat both the earning and the expenditure sides of the income issue. Case studies show that issues of ownership, access and control over income are of greater importance for women than men because women's relationship to productive resources is more vulnerable.[24] By monitoring women's expenditure patterns it should be possible to determine the extent to which data on women's income accurately reflect their full contribution to the welfare of their households and families. Women have numerous sorts of 'income' which may not be accurately reflected in standard income definitions but can be revealed through data on their expenditure commitments. For example, it should be possible to record information on women's expenditure and the same of the income used to finance it: that is, own-production, wage work or transfer from other family members. Similarly when women earn a wage income but retain little or no control over it, this would be revealed by their limited or negligible expenditure commitments. While this is not a watertight solution, not least because incomes are a highly fungible and mobile resource, it would greatly enhance the kind of data currently available to researchers and policy-makers.

Control over income

Quantitative data on per capita income do not inform policy-makers about the control or use of that income, just as data on total household income do not adequately testify to individual well-being. For example, a wife may have direct control over 'in-kind' income but little or no say over the use of her cash income.[25] Quantitative budget data can be used to test the assumption that members of a household share a common standard of living. Disaggregated by gender, the data can show differentials in expenditure on health care, education and clothing. Data from nutrition and health surveys can also be used to point to differences in the welfare of household members.

Data requirements

To obtain a real sense of women's economic contribution, it is necessary to expand the concept of income. INSTRAW has made a number of recommendations that data must include: [26]

- Individual income in cash or in kind received from all forms of recognized permanent or casual employment
- Cash and non-cash income received from own account work such as trading or working in the informal sector
- Individual and household transfers received from absent family members, other kin and/or households
- Individual and household transfers received from the state in cash or

in kind (pensions, subsidies, allowances, etc.)
- The monetary value or income equivalent of economic activities women perform for the household free of charge which would have to be paid for under other circumstances.

Failure to account for the value of home production by women and children seriously underestimates their economic contributions in developing countries. Yet measurements of home production are complicated and difficult to compare cross-culturally. At present, the sort of data needed for a detailed income and income-distribution analysis is not available in national income statistics. Moreover, it is unlikely that much progress will be made in the short- to medium-term in incorporating the value of home production and the contributions of women and children into national income estimates.

Sources such as agricultural sample surveys, village studies and specialized household or community surveys are probably more capable of handling an expanded concept of income and incorporating detailed income questions into their existing methodologies. Limited progress has been made, for example, in Farming Systems Research and detailed studies of household economic behaviour. Some analytical progress towards improving data on income has come about through the application of sophisticated household models which attempt to value time spent in household production by imputing a market value to it (equal to the current wage rate), and through the collection of quantitative and qualitative data in intensive household and community-level surveys.[27]

SOCIO-ECONOMIC INDICATORS: TOOLS FOR BETTER ANALYSIS AND PLANNING

Once data are gathered on the position of women in a country, the statistics must be analysed and interpreted. Where do the women stand in absolute terms and in relation to men? How do different development strategies affect women? And how are the women affecting these strategies?

Socio-economic indicators are extremely useful tools here. Indicators give meaning to raw data; they summarize information in a single variable and in such a way that it gives an indication of change. For example, an index of industrial production is commonly used as an indicator of economic performance. Similarly Physical Quality of Life indices are used to summarize a variety of data on social and economic well-being. Underlying each indicator is an assumption not only about what data are important but also what kinds of social and economic relationship are important. Indicators can be used to explore the relationships among variables measuring different social and economic conditions and trends. For example, a socio-economic indicator might show the relationship between a woman's level of education and the number of children she has, or the nutritional level of the household in which she lives and the income of that household, or the household's use of development-programme options offered in her community. Indicators can also be used to define policy goals and measure progress towards those goals.[28]

A number of international data-gathering agencies are now constructing national and international indicators on women. The goal is to improve the ability of policy-makers and planners to identify and diagnose development problems, so that gender-aware programmes can be designed to deal with them. However, in compiling these indicators, some general difficulties have to be overcome and planners should be aware of them.

First, the indicators need a critical mass of statistics, yet much data on the situation of women in developing countries are missing. Second, the indicators can exaggerate conceptual and methodological biases in the raw data. This can lead to spurious statistical results that are misleading about the actual situation. Third, indicators should be able to monitor trends over time, but it is difficult to compile consistent longitudinal data for indicators in most developing countries. Fourth, indicators will differ from country to country because no country has the same degree of data coverage. Population censuses are only useful for the most basic and conventional indicators.

Nevertheless, great progress has been made in compiling various sets of socio-economic indicators which can be useful in problem diagnosis, policy-making and planning for gender equity. A project commissioned by the Population Council used data for 1970–80 from 75 countries to study the interrelation between women's status, men's status and processes of development.[29] A vast array of indicators were drawn up for the exercise, focusing on four major sets of relationships:

1 The relationships between different socio-economic indicators of women's status
2 The relationships between indicators of economic and social development over the period 1970–80 and indicators of women's status in the same period

3 The relationships between women's status and differences in male–female status, and
4 The relationship between economic and social development and differences in male–female status.

Socio-economic indicators like these measure disparities and inequalities between women and men in relatively conventional conceptual areas such as education, health and employment. They also tend to treat women's status as a composite index which is assumed to refer to the majority of women. By definition composite indicators tend to smooth over differences that exist between women because they rely on aggregate data and focus on measuring only general and often over-simplified trends.

However, problems of sex differentiation also need to be addressed in the areas of access to resources, power and authority. Moreover, specific forms of inequality and discrimination usually take on different meanings for women from different social categories. Thus for indicators to transmit the correct message they should differentiate and disaggregate among groups of women. To be most useful indicators for different socio-economic groups of women need to be compared against each other as well as against groups of men.

In recognition of some of these points, a group of researchers working under the auspices of UNESCO formulated a number of ways forward.[30] They proposed that at the very minimum indicators on women should transmit information in three main areas:

- *Structural factors affecting women's participation in development,* including socio-cultural constraints affecting access to available resources, legal limitations to work and the legal status of groups of women
- *Women's contribution to development,* including the nature of and participation in market and home production as well as part-time and informal sector employment
- *The benefits women derive from employment,* including sex differentials in the provision and use of services, conditions of work measured by earnings and access to factors of production and infrastructure.

A group of Caribbean researchers has tried to develop an equity-oriented framework for indicators that goes well beyond the scope of conventional indicators. The framework emphasizes the interrelationships between economic, social and political factors and stresses the

Table 2.3 Massiah framework of indicators*

Dimension	*Sub-category*	*Data variable*
1 Resources	1.1 Human	a) Population b) Education c) Health d) Social conditions
	1.2 Physical	a) Land
	1.3 Economic	a) Patterns of ownership b) Performance of economic resources
	1.4 Social mobility	a) Socio-economic groups b) Mobility c) Equality of opportunity
2 Status of women	2.1 Sources of livelihood	a) Involvement in the production process b) Other source
	2.2 Emotional support	a) Involvement in reproductive unions b) Extent of motherhood c) Leisure activities d) Job satisfaction
	2.3 Power and authority	a) Familial b) Economic c) Social participation d) Political participation
3 Legal provisions	3.1 Legislation on sex equality	a) Equal opportunities at work b) Occupational distribution of women and men c) Marital/divorce rights

**Source*: adapted from Massiah (1981), from p. 82 of original ms.

importance of indicators measuring social mobility, legal status, cultural development and political participation.[31] This system of indicators has both quantitative and qualitative measures which indicate differences in the participation of women and men within the society, as well as differences in their perception of their own and their group's position within that society.

The Massiah framework of indicators is shown above, in Table 2.3. It has three dimensions: resources, the status of women and the legal provisions that relate to the various aspects of women's lives that are so often neglected in aggregate analyses of their status. Each dimension requires data on a number of sub-categories: human, economic, power and authority, and variables which relate more closely to the kinds of data already available or which should be available. This framework constitutes a major challenge to conventional uses of socio-economic indicators and, although the data requirements in the framework go well beyond anything currently available in developing countries, it provides a useful guide to the kinds of socio-economic indicators that are relevant for truly gender-aware diagnosis and planning.

While socio-economic indicators are an informative way of using existing data, it must be understood that they cannot be a substitute for developing new concepts, methods and allocating resources for collecting data that measure the situation of women as it truly is in each of the developing countries. Only if we know where we were and where we are now, can we take a confident step forward.

NOTES

1 Goldschmidt-Clermont (1982); Beneria (1982).
2 Boserup (1970, p. 163).
3 Deere (1982).
4 Sen and Sen (1985, WS-52).
5 Thorner and Ranadive (1985).
6 Standing (1985, WS-31).
7 INSTRAW (1984a, 1984b).
8 Baster (1981).
9 INSTRAW (1984a, 1984b).
10 INSTRAW (1984a, 1984b).
11 INSTRAW (1984a, 1984b).
12 INSTRAW (1984a, 1984b).
13 ILO (1976, p. 32).
14 Beneria (1982, p. 128).
15 See Chapter 4: *Employment*.
16 Berry and Sabot (1981).

17 ICRW (1980).
18 King and Evenson (1983).
19 Mueller (1982).
20 Dixon-Mueller (1985).
21 Anker (1980).
22 INSTRAW (1984b).
23 INSTRAW (1984b). For further elaboration on these issues, see Chapter 8: *Household resource management*.
24 INSTRAW (1984b, p. 75).
25 See Chapter 8: *Household resource management*.
26 INSTRAW (1984b).
27 Goldschmidt-Clermont (1982).
28 Baster (1981).
29 Safilios-Rothschild (1986).
30 UNESCO (1981, p. 102).
31 Massiah (1981).

Chapter 3

Agriculture

*Ann Whitehead and Helen Bloom**

Over the past twenty years people have come to realize that women play an integral role in African agriculture and that they produce a high proportion of its food.[1] But they also do more than produce food for self-consumption. Many rural women function as independent farmers as well as family-farm labourers and in many cases they are involved in the production of cash and market crops.

Yet when development agencies deal with sub-Saharan Africa, most of their policies and programmes overlook the real roles played by these women and the contribution they are ready and eager to make in solving the food crisis in their region.

It must be understood clearly: the rural women of Africa *want* to increase their agricultural production. There are real factors preventing them from doing so. So when designing policies for women farmers, we have to ask ourselves: 'What is holding them back?' There are some general answers to that question.

Some of the constraints on women's production are related to the sexual or gender division of labour – the way cultural concepts and traditions define what 'work' is, under what relationships it is performed and who does it.[2] Others are derived from women's access to resources and the effects of commoditization and development-planning on the division of labour in the farming household. Still others are based in the dual nature of women's economic roles within the farm family.

The cultural basis of the gender division of labour suggests that – like other social and economic relations – it is subject to change. The form these changes take, as the rural production systems of sub-Saharan Africa undergo economic transformation, is very important. In this chapter we look first at the gender-related factors that are causing low agricultural production among the women of sub-Saharan Africa. We then consider ways in which development planners can help women to overcome present constraints and avoid creating new ones.

GENDER ISSUES IN AGRICULTURE

In sub-Saharan Africa, to a large extent women's agricultural production is being impeded by gender issues. This occurs in three general ways:

- First, women are not being given opportunities to increase their agricultural production because sex stereotyping is affecting development-planning at local, regional, national and international levels.
- Second, because of the way gender relations organize access to resources, women's claims to resources for independent farming have diminished with commoditization and commercialization.
- Third, women's production is affected by the conflicts between male and female household members over the use of each others' labour and over their respective rights to consumption goods and income produced within the household.

SEX-STEREOTYPING IN DEVELOPMENT-PLANNING AND POLICY

One of the recurring problems for rural women in sub-Saharan Africa is that their work is invisible. It is inadequately recorded and inadequately recognized. As explained in Chapter 2: *Statistics*, Africa has not escaped the widespread problem of sex discrimination in statistics, which has been reported from all over the world, nor the effects of the inadequate recording of women's work.[3] There is a large discrepancy between the amount of women's work reported from micro-studies and case studies and that recorded in employment and agriculture macro-statistics.

Dixon-Mueller documents how statistics collected through farm and agricultural censuses enumerate two to ten times as many unpaid female workers as other sources.[4] This is because international standards are biased towards urban and developed countries' notions of work, employment and production, while the data-collection methods of farm censuses are much more in tune with the nature of rural and Third World work patterns. Although there are some exceptions – for example, the national series of Kenyan agricultural statistics – most tropical African countries have extremely few macro-statistical data sets on which to base planning decisions about rural women's work.[5]

An equally important reason behind the underrecording and economic invisibility of women is that sub-Saharan rural production is characterized by a high proportion of smallholder units in which production for sale goes on side by side with production for own-consumption. Dixon-Mueller argues that farm censuses are not very consistent about what is treated as production for own-consumption and what is not. Because women's activities in sub-Saharan Africa are weighted more towards the non-market sector than men's, this aspect of economic measurement has important gender implications.

Three other reasons why women's work is not adequately measured is that data collection depends on (i) who is interviewed, (ii) how family labour is recorded, and (iii) whether only the main or primary activity is recorded. Misconceptions at these levels are particularly relevant to sub-Saharan Africa. The economic separation of husbands and wives makes it extremely important to interview women independently of any (presumed) male head of household.

Women's work, like men's, is often the subject of cultural valuations that obscure its character. A good example is the use of the term 'garden' to describe women's farms – farms often responsible for feeding an entire family! The lack of significance that rural culture and development planners attach to women's family labour is one of the main causes behind the mismeasurement and underenumeration of women as members of the agricultural workforce.[6] Women's family labour is often viewed as an essential part of the obligations of the wife, mother and daughter – a continuation of the woman's social roles rather than real 'work'. When men do the same tasks, it may be labelled economic activity. Such subjective assessments condemn a high proportion of the potential rural workforce to being overlooked by development planners and policy-makers, a situation that can only hamper any possible success.

Yet sex stereotyping and cultural evaluations often exist within development-planning and policy, where they operate over a wide social field. For example, most economists view subsistence production as 'non-market production' which revolves around the 'family farm' and is taken care of by 'domestic relations'.

Emphasizing the domestic character of self-provisioning allows a number of ideas to be conflated and mistakenly fused:

- There is an assumption that self-provisioning does not need to be planned for, but as part of domestic life can be left to take care of itself.

- Using the term 'domestic' accentuates the idea that family mores are at work.
- The term 'domestic' can signal that because women are bound into the family and have a strong interest in their children, the women will be less innovative, less commercial and more risk-averse in their dealings outside the home.

This can be seen when we examine the basis of the 'invisibility' of women's work which leads to the widely publicized problem of labour bottlenecks. Micro-economic agricultural planning in sub-Saharan Africa has a lamentable record of introducing labour-demanding crop innovations which clash with the peak demands for women's labour. Underlying this erroneous planning is not only the lack of visibility of women's work, but also a lack of understanding on the part of agricultural economists of the economic value of a wide range of non-market produced goods and services which the farm household cannot forego.

We must recognize, however, that there are some real differences between men's and women's work in sub-Saharan Africa. For a number of reasons, in this region it is impossible to consider women's productive work in isolation from their reproductive work. Thus a woman's productive tasks may need to be performed close to home, to withstand fragmented attention and to have a shifting and fluid pattern so she can adapt her scheduling according to changing priorities. Nevertheless, productive and reproductive work are *different kinds of activities*: they are affected by the market quite differently. We need a standard definitional boundary between them so we can begin to collect true information about this sector of the workforce.

Extension services

Another main finding in the literature is that there is widespread discrimination against women in extension services and agricultural innovation. The most systematic study of the effects of extension services on rural women was carried out by Kathleen Staudt in Western Kenya.[7] She noted, among other things, that in a district which had little contact with extension workers, the productivity of men and women was about equal; but in a district with extension workers the relative productivity of women had declined. Examining samples of farms which were similar in all respects except the gender of their heads, she found farms with male heads were four times more likely to have been visited by an extension worker. Staudt also found that 'capable women were being ignored for non-innovative men'. In a similar vein, Fortmann noted that in Tanzania 'the conventional wisdom that women cannot reason as well as men reduces any incentive for working with women'.[8]

Other, more general studies confirm these detailed findings:

- Women have few agricultural extension services directed at them, but are mainly the recipients of home economics extension.[9]
- Very few women farmers are contacted by agricultural extension workers.[10]
- The staff of agricultural extension services are overwhelmingly male.[11]

This lack of attention is due to discrimination; it is not because women are inferior farmers. Different studies show that women farmers can be as productive, as efficient and as modern as their male counterparts.[12]

Although such problems may arise from men and women in the rural culture, the most significant point is the strong bias towards men in the state agricultural services. The history of this neglect of women farmers can be traced back to mission and colonial times, and, in most places unfortunately, continues in the post-colonial period.[13] The discrimination systematically blocks women's access to critical knowledge and inputs which could help them improve their productivity. In so far as the education and training of extension workers actively incorporates sex stereotyping, so it could also seek to combat it.

RESOURCE CONSTRAINTS

It must never be *assumed* that the primary reasons why women do not innovate or produce more are because of the effects of cultural or religious values about their proper behaviour, because men will not let them or because they are too 'overburdened'. Rather, we must ask whether there are any resource constraints which act as a material basis for this state of affairs. For example, a 1975 USAID review of the position of women confidently states that women do not plough in North Ghana because there is a 'taboo' which forbids them to touch cattle. Yet research among the Kusasi group, which has the largest take-up of ploughs in the region, suggest that even if this 'taboo' existed, the real reason women do not plough has more to do with the predominant position of men in the millet farming system and the almost exclusive male ownership of cattle. The women have no access to beasts to pull the ploughs.[14]

Time constraints

The theme of women's enormously heavy work burdens is very dominant in many policy documents. A number of studies have stressed that women cannot do more productive work because they simply do not have enough time: they are overburdened by the combination of productive work and domestic reproductive work, especially in conditions of environmental degradation. Yet this viewpoint is not unanimous. The evidence of women's heavy work burdens and its implications on agricultural production need to be interpreted very carefully.

Because women carry a double workload, the demands on their time are always greater than on men's and they often work longer hours. On the whole, women's work burdens have *increased* with economic change. Kitching argues that the main burden of increasing agricultural production between 1900 and 1945 in Kenya lay on women's shoulders and similar processes are visible elsewhere.[15]

However, the evidence that these work burdens are 'intolerable' in the sense that women's scarce labour time is the major constraint on certain forms of increased productivity or agricultural innovation is much less clear cut. Many projects and studies show that women's high work burdens can be increased still further if the women can be directly or indirectly coerced, or if they can be motivated by economic incentives directed towards themselves.

Access to land, labour and capital

However, the most important resources to consider are those which are critical to farming, namely land, labour and capital. Here it is important to treat the household's access to resources in a disaggregated way. Although this is the case in relation to women's resources in all economies, it is particularly important in sub-Saharan Africa because of the particular character of women's economic roles within the family.

THE SEXUAL DIVISION OF LABOUR

The importance of rural African women's productive roles was first made clear by Esther Boserup in an influential and pioneering study.[16] She argued that sub-Saharan African farming systems could be described as female because of the importance of women's labour input. Moreover, she maintained that during the twentieth century 'modernization' produced a dichotomy in which men work in a sector of enhanced productivity, either in farming for the market (which might include plantations, commercial farms or the peasant sector) or in migrant urban and industrial labour, while women remain in an untransformed sector in which they use traditional and low-productivity technology to farm for subsistence.

Boserup's findings were very significant in bringing the enormous role African women play in rural production to the forefront of public thinking on development. However, in their popular forms they have often been oversimplified, for example by suggesting that African women produce most of the continent's food, especially that produced for subsistence. It is more accurate to point out that women *and* men are involved in a complex gender division of labour in agriculture.[17] In many areas men grow certain kinds of food crops and women others. Women's food crops are not only produced for consumption – they can be, and are, used as cash crops too. In some areas women have led the development of the production of cash crops, especially where male labour was engaged in other ways in the modern economy.

This gender division of labour is situated within a complex set of rights and obligations within the family. There are two critical features of these family forms which have affected how women experience economic change and development.

First, this gender division implies an interdependence between family members in both the productive and the reproductive spheres, creating the need for an exchange of goods and services. But there is no *a priori* reason to assume this exchange takes the form of sharing and is intrinsically harmonious – an assumption planners conventionally make. This notion of sharing is more appropriate to some Western family forms[18] than to rural Africa, where many researchers have noted a striking separation of the domestic budgets of men and women.

Second, the essence of women's productive role in the contemporary African farm family is that they have dual work situations. They have access to land and other resources for independent farming, but they also work as unremunerated family labour on the fields of male household members. This dual role implies very different conditions of work for each kind of activity.

Women's dual economic role in the family economy

The roots of women's dual role lie in the forms of the obligations which existed between men and women in the societies and economies of the nineteenth century. There was a whole variety of sub-Saharan societies then, and these could be described as lying on a continuum of economic development. At one end there were the closed economies, which were based on kinship collectivities producing entirely for self-provisioning and internal markets. At the other end of the continuum there were complex forms of pre-industrial state, in which socio-economic differentiation, usually involving slavery and considerable production for exchange, had developed.

Slavery and forms of social stratification played an important part in those societies with more complex divisions of labour. Domestic and conjugal relations – plus kinship relations in the widest sense – were important in the economies of *all* societies of this period. There were four important features of the relations between the sexes:

1 Access to resources was unequal. Women's resource rights in the pre-commodity economy were based on their position within the kinship structure and the household. Their rights were guaranteed by the forms of authority and power within these systems, but they were very different from – and often much less than – men's. This was most marked in those societies where the accumulation of wealth had occurred, for example in bridewealth societies where women never had the same rights to bridewealth livestock as men.
2 The resources of husband and wife were not merged. No single joint fund or common conjugal property was established on marriage.[19] In other words, there was an economic separation between husband and wife.
3 Responsibility for providing for the well-being of children was divided between the father and mother. Societies differed in the extent to which providing food for the children was the mother's responsibility, although often it was solely hers. Some people have suggested that this was because it was common for a man to have more than one wife. Whatever the reason, responsibility for the well-being of the children required women to be economically independent to some degree.
4 Husbands' rights over their wives' labour were generally extensive and sometimes entailed direct remuneration of the wives. Marriage usually obliged women to work for the households and kin groups into which they had married. A wife's rights to use the labour of the male household members were generally less extensive, although a husband often had considerable labour obligations to his in-laws.

The twentieth-century economic revolution in sub-Saharan Africa has transformed the pre-colonial family role of women into a very different, modern form of the dual role. During this historical process, rural work burdens in the African smallholder sector have increased. Male labour has seeped out of the subsistence sector into urban wage work or into agricultural employment with the development of commercial cash crops. Thus it became more important for women to pursue their independent farming, but it became increasingly difficult for them to do so. As a result, women's agricultural work burdens have increased with relatively little increase in productivity.

Some of the causes for this are found in the many structural and institutional features that affect women's use of the available land, labour and input resources.

In their work as independent farmers mainly growing food crops, women are finding it very difficult to get the resources they need: they lose out against men in the competition for land, for labour and for improved agricultural resources. As more land is taken into production and as it becomes scarce, women have had difficulty protecting their land rights on the basis of either local or state codified procedures and laws.

Numerous studies show that, in general, women have been at a disadvantage under the changes in legal ownership and under the changes in social relations which give access to land.[20] For example, matrilineal inheritance is no guarantee that women can claim land when title is being registered.[21] Some state legislation, as in Tanzania, has discriminated by sex.[22] In many cases women have no leverage in the network of political links that controls the *de facto* distribution of rural land rights.[23] Moreover, resettlement and irrigation schemes which require large changes in land use *somehow* manage to fall most often on women's land, rather than men's.[24] Thus the resource base for women's independent farming is undermined.

At the same time as these changes have been occurring, wives' labour has become relatively more important within the total family labour supply. An increasing proportion of women's labour time has been spent in production for their husbands. When African households' cash requirements were increased by colonial rule – either directly by tax demands or indirectly by new consumption goods – the main immediate avenues for earning such income were men's cash-cropping or migrant labour. Women members of the household were able to make their contribution to increased cash needs by their work as family labour in cash-cropping or by increased trading. In the initial phases of these processes, in so far as women's welfare was bound up with that of their households, there were simple incentives for the women to do this work.

But over time these decisions to undertake more work as 'family labourers' have taken on new economic meanings. As the terms of trade declined for peasants, as land became scarce and as rural differentiation proceeded, there was increasing evidence of acute stresses and strains. There is no guarantee that the increase in labour time now required for peasant domestic and productive activities affects men and women equally. The potential for coercion within the customary obligations of a married woman to her husband may become an important element in her increased workload.

In general, women have much greater control over income from their own independent farm work – how they spend the income and who benefits from it. It is very much harder for them to determine the spending patterns of the 'household' income from their work as family labour. Different spending preferences between husbands and wives, including different assessments of the importance of children's welfare, reduce the incentives for women to do this family labour. There is increasing evidence that some wives are resisting increased work on their husband's fields because of the welfare effects on themselves and their children. Yet even today most development projects write a woman's labour in as family labour, without checking whether her work will be available and what the costs are of the woman foregoing her work in other areas.

Women's capacity to farm independently is also being eroded by their changing access to other people's labour. Male migration erases the contribution of male labour which women used to have through their conjugal and domestic relations. To compensate for the effects of migration, men use their social and political networks to arrange work-party exchanges, but women have no male labour to trade. In addition, women tend to lack the money resources to hire labour.

Capital

A number of studies have argued that women's ability to respond to new economic opportunities is being constrained by their initial comparative lack of overall resources. Stored wealth – in the form of livestock, machinery, etc. – is more often in the hands of men. Many rural women do not control enough cash to be able to hire ploughs or buy seeds, fertilizer or new technology. Thus when agricultural development projects require participants to purchase inputs and set up enabling credit schemes, most female farmers find it impossible to fulfil the conditions of the credit schemes. Planners have unwittingly created a formidable barrier to women's access to new inputs.

It is important to understand that all the members of a household need both food production for self-consumption and cash-cropping. Many items of household consumption must necessarily be bought. Withdrawal from the market would imply a deplorable level of unmet basic needs. The main problem in sub-Saharan Africa is not that cash-cropping produces rural starvation, nor that food production is in the hands of underresourced women. Rather it is the competition over which crops should be allocated the scarce resources of land, labour, fertilizer and other inputs. Men and women experience this competition differently because of their different economic roles within the family.[25]

Poor women and female-headed households

Another important change that has occurred in sub-Saharan Africa is that the combination of labour migration, investment in agriculture and increased commodity production has produced higher incomes for some rural households, change in kinship relationships and new kinds of socio-economic differentiation. Women have benefited most in the commercial farming and trading sectors, where they have enhanced opportunities for increased agricultural income, trade and other forms of production like beer-brewing. But the rural areas have seen the simultaneous development of a stratum of female-headed households which lacks the resources to meet production and consumption needs. A growing body of literature stresses that rural female-headed households are often very poor indeed and that their numbers are growing. Many of them are no longer able to farm independently and subsist on casual labour paid in cash or kind.[26]

CONFLICTS WITHIN THE HOUSEHOLD

As we have described, prior to colonialization, rural households in sub-Saharan Africa had a relatively complex structure of interlinked rights and obligations which served to simplify the economic implications of polygamy and tied the economic activities of the household to those of the wider kin group. Women's dual economic role, which required them to work for their husbands or other male members of the family as well as perform independent economic activities, was part of that system.

Many of the changes taking place in the rural economy and in the gender division of labour in the production, distribution and accumulation processes are linked to radical changes in the economic relations of households and kin groups. It is therefore not surprising that one of the most widespread and persistent findings is that changes in the conditions of work of rural women are setting off wide-ranging domestic conflicts between men and women. These include struggles over the gender definition of tasks (which crops must women weed?); over the disposition of household labour (e.g. the labour input to cash-cropping); over the distribution of household income; and over the patterns of household spending and consumption.

One result of commercialization on African agriculture is that women are spending an increasing amount and proportion of their labour time on their husbands' farms. When colonial rule increased the African household's cash needs – either directly (by tax demands) or indirectly (by new consumption goods) – men's main avenue for earning the cash was via cash-cropping or migrant labour. Women's contribution was via increased trading or unpaid labour on the cash crops.

Roberts argues that as a region moves towards the market system and as the bonds securing other unpaid family labour (e.g. sons) are destroyed, wives' labour becomes more important to their husbands.[27] This labour is rarely paid directly and it can only be released by divorce. It is important to note that some researchers believe the potential for coercion within the system is increasing. Roberts' research suggests that the struggle between the sexes over women's labour has been very important in the historical development of market production in West Africa.

To women the increased time spent on husbands' farms is very important because they lack control over the rewards for what they produce. Rarely is there a direct link between the amount of a wife's family labour and the proportion of the total household product or income to which she has access. The pre-colonial systems of distribution of rewards (which may not have functioned equitably in the first place) do not automatically lead to sharing when the household product is in the form of money.[28]

For example, a very early study [29] of a large MWEA irrigation and resettlement project in Kenya showed that women and children suffered because of lack of control over an increased money income. Because the project made much less land and labour available for food production, the planners proposed that a certain amount of money be taken from the increased income to be spent on food. However, women were not in a position to control the allocation of the money, so they and their children suffered. Project implementation often brings about a decline in nutritional status because of gender differences over income allocation.[30]

DEVELOPMENT-PLANNING AND CONFLICTS IN MARRIAGE

Despite the fact that the dual form of women's traditional economic obligations in sub-Saharan Africa is well documented, it is an almost universal rule that development planners assume wives are available to work on husbands' crops as unpaid family labour. Virtually no development projects include the expansion of women's independent farming.

The model of the household economy usually used is based on a married couple and other family members where all work and share together. The husband/father is regarded as managing the resources on behalf of the household, inputs are channelled to the farming enterprise through him, while the others are regarded as 'his dependants' who provide labour under his direction. The widespread use of this model has created innumerable conflicts and problems.

For example, an attempt to introduce irrigated rice production in The Gambia [31] made an initial assumption that the men were the traditional rice growers and had full control over the resources required for it. In reality women grew the rice for household consumption and exchange within a complex set of rights and obligations between husbands and wives. Backed by project officials, the men established exclusive rights to the women's rice fields and pushed the female rice farmers out to inferior scattered plots to continue cultivating traditional rice varieties. The levels of improved rice production were disappointingly low, partly because women were reluctant to participate as family labour (their planned role) and husbands had to pay them to work on the irrigated rice fields.

Sometimes the conflicts caused by the use of this erroneous conjugal model threaten the entire success of a development project. For example, Tobbison describes how women farmers, who predominate in food-crop production, resisted a programme to encourage cultivation of hybrid maize by the distribution of subsidized seed, fertilizers and pesticides to men. The women protested because their increased workload brought no concomitant control over the income into the farm household.

CONCLUSION: DEVELOPMENT-PLANNING AND GENDER ISSUES IN RURAL PRODUCTION

There is a persistent belief throughout sub-Saharan Africa that development projects, far from helping women, appear to make them poorer or to affect them adversely in other ways. A growing and sophisticated body of literature is making a determined attempt to examine the reasons for this belief.

Evaluations of agricultural development projects in terms of their impact on rural women have found a set of linked problems which are central to why projects fail the women:

1 Development planners have difficulties in comprehending the economics of the partially self-provisioning farm enterprise and particularly the role played by women's work within it.
2 When designing projects, planners frequently fail to recognize the extent of women's work in agricultural production. As a result, male farmers are targeted for inputs and extension work on crops grown only by women.
3 Planners often fail to appreciate that innovative agricultural practices can affect other kinds of 'invisible' female labour. For example, subsistence or food-crop production may become much more difficult and/or less rewarding.
4 Planners neglect the economic significance to household welfare of women's productivity and effectiveness in reproductive and off-farm activities.
5 The model of the social relations of the sub-Saharan family farm enterprise used by development projects does not fit the complex form that these social relations actually take. In particular, as described above, planners repeatedly assume that men can use women as unpaid family labour; this leads to intense domestic conflicts between men and women.

Development policy-makers and planners are very sensitive to the issue of whether rural men will support projects for women. Evaluations by Buvinic and by Dixon[32] show that male support is a crucial variable in the success of projects for women. It appears that male opinion leaders in the community are more likely to think in terms of welfare and health projects for women, in much the same way as government officials in receiving and donor nations. There have been clear instances when male opposition has sabotaged efforts directed at rural women's economic well-being.

The reactions of rural men to development-planning is obviously a factor to be dealt with, but it is one among many. Many policy-makers in donor nations express timidity and reluctance about interfering in the social relations between men and women. A distinct sense is sometimes given that planners fear a crowd of irate husbands sabotaging their efforts to improve the wives' lot.

Interpersonal relations between men and women are often thought not to be the proper subject of planning. Yet evidence from numerous studies suggests that what is already being promoted and prosecuted by development is resulting in an enormous upheaval in the domestic relations of men and women and has had profound effects on gender relations. Many projects implicitly construct new roles for women by their Eurocentrism, or by their neglect of women. Such changes only deepen women's disadvantages and their women's ability to contribute to the improved welfare of their families and regions.

To avoid these errors, development experts should bear in mind two guiding principles when dealing with sub-Saharan Africa. The first is to emphasize constantly just how alien the domestic structures are which much planning assumes and just how much interference they create in the interpersonal relations of the sexes. The second guiding principle is to emphasize that gender issues affect both values and practices. To have a beneficial effect, planners must seek to obtain *gender-disaggregated analyses* of the social and economic aspects of the rural production system, and then try to introduce solutions objectively.

NOTES

* This chapter is adapted from the training manual *Gender-Aware Planning in Agricultural Production*, but a fuller account is to be found in Whitehead (1990a).

1 FAO (1984, p. 1).

2 See pp. 9–10 in Chapter 1: *Gender*, for a more detailed discussion of the gender division of labour.

3 See UN (1984); IBRD (1979); ICRW (1980); Boulding (1983); Beneria (1981, 1982).

4 Dixon (1982); Dixon-Mueller (1985).

5 See the discussion on pp. 30–5 in Chapter 2: *Statistics*, for a fuller discussion of the absence of rural women and their work from official statistics.

6 Dixon-Mueller (1985).

7 Staudt (1985).

8 Fortmann (1981).

9 Feldman (1981); Bryson (1980); Jackson (1985).

10 Fortmann (1981); Gaobepe and Mwenda (1980); Staudt (1978).
11 Safilios-Rothschild (1985); Fortmann (1981).
12 Moock (1976); Fortmann (1981); Staudt (1979).
13 Muntemba (1982).
14 Whitehead (1981).
15 Kitching (1980).
16 Boserup (1970).
17 For more complex views and critiques of Boserup, see especially Guyer (1983) and Richards (1984).
18 Although even for these it may also be an unwarranted assumption. See references in Chapter 8: *Household resource management*.
19 There is a wide-ranging literature discussing this economic separation and its implications. See Guyer (1983); Goody (1976); and for bibliographies see Guyer (1981) and Whitehead (1984).
20 e.g. Caplan (1983); Muntemba (1982).
21 Mascarenhas and Mnilinyi (1983).
22 Muntemba (1982).
23 Whitehead (1984); Clark (1981).
24 Cloud (1976); Conti (1979); Jackson (1985); Hanger and Moris (1973); Dey (1981) and the discussion in Palmer (1985).
25 This aspect of changing gender relations is discussed more fully in Whitehead (1990b and 1991).
26 This chapter has concentrated on the problems facing women farmers in smallholder households. The problems facing the many women working as agricultural wage labourers and female heads of households are currently under-researched and demand greater attention. See Yousseff and Hetler (1984).
27 Roberts (1983).
28 See Chapter 8: *Household management*.
29 Hanger and Moris (1973).
30 See Tobbison (1984).
31 Dey (1981, 1982).
32 Buvinic (1984); Dixon (1980).

Chapter 4

Employment

Hilary Standing

This chapter aims to demonstrate the importance of gender in understanding patterns of employment. With the aid of case-study material from India, we will show how awareness of gender issues can influence policy initiatives to improve both the quality and the quantity of women's participation in the workplace.

In doing this our approach focuses on gender rather than on women *per se* for several reasons. First, the terms 'men' and 'women' may signify quite different things in different social contexts and also vary across classes. For instance, women of higher classes may be associated with gentleness and passivity, while labouring women may be associated with the opposite characteristics. Such gender attributions may be used to justify women's exclusion from or incorporation into different kinds of paid work. Similarly, common-sense accounts of the sexual division of labour into heavy ('male') tasks and light ('female') ones – corresponding to an objective difference in male and female strength – are contradicted by the many cases in which women carry out those same 'heavy' tasks in their other roles in society and by the cultural variations across countries.

Second, the highly uneven distribution of the sexes within all sectors of the labour force, and the tendency in most countries for women to be located in lower paid and less skilled jobs, suggests that gender is a major variable in determining labour-market placement. Explanations for women's relative disadvantage in the labour market cannot be found simply by looking at the structure of the labour market itself or by considering women in isolation. This is because women compete for employment on terms that are set by the wider social relations within which they operate.

These terms may be dictated by an *economic* agenda, such as the availability of child care for women who are mothers; or a cultural agenda such as strongly held beliefs about the unsuitability of employment for certain kinds of women or even women in general. Or they may be determined partly by wider political considerations in which the prior right of men to (scarce) employment is upheld.

Third, the attribution of particular characteristics to the genders also permeates sectors, and jobs, through processes of sex-typing. These produce hierarchies of gender-related skills, with women concentrated predominantly in 'unskilled' and 'semi-skilled' jobs and tasks, as well as gender segregation, which results in wholly female or male enclaves of employment. Again, the determinants of these processes must be sought outside the immediate arena of labour markets.

Fourth, a major reason for making gender rather than 'women' the unit of analysis are differences between women because of their class, race, cultural and life-cycle experiences. Although all women may experience disadvantage relative to men of the same class, not all women are absolutely deprived in terms of their access to a basic minimum livelihood. The significance of gender in the production of deprivation is thus not necessarily the same for all women at all times, nor are all women affected in the same way by social and economic change. This can have important implications for formulating gender-aware policy. For example, the introduction of rice-hulling technology in Bangladesh had beneficial consequences for women from land-holding households dependent entirely upon family labour because it relieved the women from time-consuming daily drudgery. However, it had adverse affects on poor landless women for whom paddy-husking was one of the few sources of locally available paid labour.[1]

In analysing employment patterns and policy alternatives, it is important to look at the kinds of factors noted above, such as the attribution of specific characteristics to women and men, as well as the various constraints that affect their relative capacities to enter and compete in the labour market. However, it is important not to prejudge what these characteristics and constraints are. Part of the value of a gender analysis is that it provides a conceptual framework for examining what are often substantial cultural and historical differences in the experience of women and men.

KEY CHARACTERISTICS OF WOMEN'S POSITION IN THE LABOUR MARKET

The profile of women's employment in India reveals a number of structural inequalities based on gender, which are also common in First

World economies. First, however, it must be pointed out that the terms 'employment' and 'work' are used interchangeably, yet 'employment' is not synonymous with 'work'. In full market economies, most models of employment assume a concept of gainful economic activity which is measured in market terms. This creates difficulties in incorporating into economic models non-market economic activity such as agricultural production for family consumption, and in measuring the real extent of women's involvement in productive work of different kinds. Official census and other economic data on labour-force participation rates therefore have to be interpreted with a gender-sensitive eye to their limitations. This is even more true in Third World economies, where much of women's productive work is disguised as 'housework' or unremunerated family labour and is thus not measured as employment by standard definitions.

In this chapter we have used the term 'employment ' to mean paid work. In the majority of cases, payment is monetary, but some remuneration in kind continues, notably in agricultural labour. Self-employment, which raises acute problems of detection and definition, has been excluded (see Further Reading).

Informal and formal sector employment

Further difficulties of measurement and interpretation arise because of the pervasiveness of women's employment in the informal or unorganized sector in many countries. Not only is that part of the economy largely unenumerated in official statistics, but it is usually also unregulated by or exempt from labour legislation. Thus, its extent can only be the subject of speculation. Unorganized sector employment tends to be characterized by casualized forms of employment, low entry costs, extremely low levels of remuneration and poor conditions of work.

Micro-level case studies show that women are often represented disproportionately in this sector. It has been estimated in India, for instance, that only 6 per cent of all women workers are employed in the organized sector. This suggests the likelihood of high levels of impoverishment among households dependent solely or predominantly on women's incomes.

Structural inequalities

In much of the developed and developing world, women's employment is constrained by a number of general structural inequalities[2] which can be summarized as follows:

1 Women earn, on average, lower wage rates than men and have fewer hours of paid work. Where part-time work is institutionalized, women are disproportionately part-time workers.
2 There is occupational and task segregation in the labour market which leads to enclaves of female-only employment. In developing countries women are concentrated in primary sector employment (agriculture) or in the informal sector. Their predominance in low value-added industries and services is linked to their wage disadvantage. This gender segregation in the labour market helps to maintain gender-based wage differentials.
3 Women are less formally skilled and the skills they are credited with, such as dexterity and greater patience in performing complex and highly routine tasks, for instance in electronics' assembly, tend to be undervalued in terms of wage rates. The sex stereotyping of jobs and tasks often leads to the designation of women's jobs as unskilled and semi-skilled simply because they are performed by women.[3]
4 As a consequence of the strength of 'male breadwinner' ideologies, women are persistently undervalued as wage workers. Women's entry into paid work can be discouraged by measures taken by male dominated trade unions to exclude women from male enclaves of employment, by government fiscal and social security measures, or by popular conceptions of women's rightful place as being in the home.

These general characteristics, and others, are brought to light in the next two sections through the analysis of trends in women's employment in India.

INTERPRETING SECTORIAL TRENDS IN WOMEN'S EMPLOYMENT IN INDIA

Indian economic strategy

India is the world's seventh largest country with a population of around 750 million people. It is also one of the poorest, ranking 66th out of the group of 77 developing countries. It is an extremely diverse country in terms of human and natural resources and is endowed with sufficient resources to permit a high level of industrialization. Since independence, it has pursued a policy of self-reliant industrialization in which both private capital and public-sector ownership of key industries are combined. The cornerstone of Indian long-term economic planning has been import substitution, particularly in the production of capital and intermediate goods. The pattern of industrial development has produced both a diverse and in part a sophisticated industrial sector, and chronically high levels of unemployment and underemployment. Manufacturing industry accounts for only 11 per cent of India's total employment, a figure which has remained constant since the 1960s. Despite the existence of a large 'cottage industry' and informal sector, the bulk of the population continues to depend on agriculture, and particularly on agricultural labour, for its means of survival. The pattern of agricultural development has been highly uneven, with rapid increases in productivity in some areas and stagnation in others.

India's pattern of growth has produced low levels of labour absorption and this has affected the extent of women's participation in the labour market. Unlike many southeast Asian countries, there is no significant female labour force located in the new export oriented industries in India.[4] However, this may change with the current emphasis on the liberalization of the economy.

Changes in the sexual composition of the labour force

In India, census data suggest that women's labour-force participation rates have undergone a *secular decline* throughout the century, and most dramatically between 1961 and 1971. However, in using Indian census data, planners and policy-makers must be aware of their limitations for interpreting patterns and trends in women's employment. This is because census data tend to be biased and underestimate the extent of women's participation in the labour force. There are several conceptual reasons for this underrenumeration.

- Definitions can cause underrenumeration if the criteria adopted are inappropriately narrow. For example, if employment is defined on the basis of an eight-hour working day, large amounts of casual, seasonal and part-time employment will be missed. Female labour is heavily concentrated in these categories.
- The definition of economic activity may change from one census to another. This can cause problems of comparability and have different effects on the recording of female and male activity rates.
- There are also likely to be inter-sectoral biases. Statistics on organized sector employment for both sexes are generally more

likely to be reliable and less likely to be subject to gender bias. However, undercounting is thought to be serious in the unorganized sectors and known to be severe in agricultural wage labour. Both these sectors have a high proportion of women.

There are also methodological biases which arise in the collection of census data and which result in the underenumeration of female productive activity:[5]

- The ordering of questions
- Short reference periods
- Cultural and ideological biases on the part of (mostly male) enumerators and participants.

Despite all of these factors leading to underenumeration, none the less, in India it is generally concluded that real losses have occurred in women's share of the labour market. To understand where these have taken place, and the possible reasons for the losses, the trends are disaggregated sectorally in the next section.

Agriculture

Agriculture provides some form of employment or livelihood to 70 per cent of India's workforce. For women, agriculture is a particularly important source of employment – around 80 per cent of women who are recognized as economically active are found in agriculture.

In terms of trends, between 1961 and 1971, the number of female cultivators dropped from 33 million to 16 million, while the number of female agricultural labourers rose from 14 to 20 million – a net loss of 11 million females in this sector. During the same period, the number of male cultivators rose from 66 to 70 million and the number of male agricultural labourers from 17 to 32 million – a net gain of 19 million males. These trends continued in the same directions in the 1981 census. Taking the overall increase in the population between 1961 and 1971, it is the apparent large fall in the number of female cultivators which is responsible for the major decline in overall female labour-force participation rates noted between 1961 and 1971. The rise in the number of female agricultural labourers suggests that processes other than just conceptual underenumeration are involved. These processes may be better understood by differentiating the impact of change on women by class, region and culture.

Differentiation by class, region and culture

Some researchers suggest that processes at work in agricultural development affect men and women differently. For example, Mies (1980) argues that the growth of capitalist agriculture causes the displacement of small cultivators. This causes a greater degree of pauperization among women than men because the displaced men are better placed to migrate for work, whereas the displaced women are marginalized in the labour market.

Chatterji points out that it is necessary to distinguish not only between men and women, but also between different groups of women.[6] Due to the uneven agricultural development (from one region to another), there are considerable regional variations in the way women's employment is affected. In some areas, evidence suggests that far from being marginalized, women are in fact increasing their share of labour due to their impoverishment. In the more prosperous areas, there is 'disappearance' of women from agricultural production as a mark of increased wealth and status and the ability to hire male migrant labour from poorer regions.[7] Thus class and regional differences have been widening as a consequence of agricultural development.

Chatterji has further disaggregated the data in terms of female and male specific tasks. She is able to show that an increase in female labour days has occurred randomly rather than in relation to an increase in demand for the tasks traditionally stereotyped as female. The growth in the female agricultural labour force has taken place largely among low-caste and tribal women who are the most impoverished and disadvantaged groups in the population. They have been increasingly employed in formerly 'male ' tasks as a direct consequence of their cheapness. The resulting increased competition between men and women for jobs has led to a decrease in male–female wage differentials – not by raising female wages, but by depressing male wages. The sexual division of labour in agriculture is thus weakening with the growth of wage labour.

There are marked cultural differences between communities and sub-castes which affect female participation in different ways. It would be wrong therefore to assume that all women face the same constraints or that all poor women are automatically available for labour. Dyson and Moore point to a complex of kinship and marriage practices which seem to correlate with the differing participation ratios in the north and the south.[8] In the north cultural restrictions tend to be stronger. In eastern India tribal and scheduled caste women have always worked outside the home. Higher-caste and Muslim women are often confined to the homestead irrespective of the economic condition of the household.

Manufacturing

In the early phases of industrialization in India, women formed a major part of the factory labour force. They were particularly employed in the textiles, jute, food-processing and agro-industries, as well as in open-cast mining and in plantation industries, and to some extent in the wood, rubber and ceramics industries. Jobs within these industries were generally sex segregated and tasks sex specific.

Apart from tea plantations, women's share of jobs in the older, organized sector has fallen considerably since the early part of the century. Nevertheless, two traditional industries, textiles and food processing, continue to account for about 75 per cent of female labour, just as they did in 1911.

In the post-independence years, with the acceleration of the shift from household to factory-based production, jobs in the organized sector have gone disproportionately to men. Emphasis on building up heavy industry has meant that growth in factory employment has been slow. Jobs in the more capital intensive industries are considered more suitable for men due to gender stereotyping, despite women's ubiquitous presence on construction sites and their history of involvement in early factory production.

The above statistical picture is not complete because much of the unorganized sector employment is absent from the statistics. Micro-level studies suggest that while formal sector employment for women has declined both in absolute and relative terms, informal sector and household industry employment has increased. Thus there has been a displacement of organized sector women workers into the unorganized sector, along with the entry of a new female labour force of young single women.

Another trend has been the casualization of the female workforce within the organized sector. For example, in the jute industry a large number of women are now hired on a day-to-day basis. These jobs are likely to be undercounted. In addition, there is almost certainly a significant underrenumeration of both male and female workers in household based industry (recorded as only 4 per cent for both in 1971), such as craft and artisanal family enterprises and piece-rate tailoring. It has been estimated that as many as four-fifths of female workers in textiles and food processing are in household based industry.

The processes by which women gradually lost jobs in factories, but then at a much later stage began to gain employment in other newer industries, are complex ones. The explanations most commonly put forward are first, the restrictive effects of protective labour legislation and second, technological change and rationalization.

Labour legislation

Labour legislation, first instituted in the 1920s, has been blamed for the major retrenchment of female workers and their replacement by men in previously female enclaves. Female employees in the organized sector have statutory rights to maternity benefits and are covered by a range of restrictions on their working hours (e.g. a ban on night shifts in textiles) and on the carrying of loads (e.g. in the jute industry). However, planners ought to beware of this argument. First, the 'cost' of protective legislation was generally invoked as a reason for laying-off women at points where a major rationalization of production was underway and a slimming of the workforce was required. Second, analysis of the real costs incurred by employers in organized industries as a consequence of such legislation does not bear out the argument, given the low rate of take-up of maternity benefits and the lower wage rates paid to women.

Technological change and rationalization

Technological change can affect women in a number of ways. It may reduce in toto the number of jobs for both men and women. Depending on the gender divisions across industries and within work processes, it may cause the disappearance of predominantly women's jobs. It may also lead to the masculinization of what were formerly women's jobs when, for instance, the men assert a stronger claim to what employment remains after rationalization, or when technological 'up-grading' of jobs causes men to be regarded as more suitable. In other cases, 'feminization' of previously male jobs may occur due to down-grading of some tasks. All these trends can be found in the history of Indian manufacturing and therefore it is important to pay attention to the gender typing of jobs and tasks in considering the impact of technological change.

Case study: jute

An example of an industry where both the technological rationalization and the masculinization of women's jobs have occurred is that of jute. In 1921, nearly one-fifth of its workforce was female (45,000), while in 1972, less than 7,000 women workers remained, although the total strength of the workforce remained more or less constant. Previously, there used to be a strict division of work processes along gender lines and women were traditionally employed in preparation work and hand sewing. Some of these processes were early targets of mechanization and women's jobs did disappear. However, many jobs formerly done by women were simply taken over by men. Rising levels of male unemployment increased the competition for jobs and employers, male workers and their trade unions all accepted that men, as breadwinners, had a prior right to employment. Ideological factors have thus played a significant role in the reduction of the female manufacturing workforce.[9]

The declining participation rate in the older industries has only been partially reversed by the growth of female employment in the newer industries. Much of this is in the unorganized sector and the workforce is completely casualized.

Case study: new employment

In Calcutta and other major cities an unknown number of women, characterized as semi-skilled or unskilled, assemble chemical, electrical and electronic components in their home or in roadside sheds on behalf of household-name companies. Female labour is preferred because it is less militant and cheaper. In such cases, therefore, new technologies appear to favour women, but at a high cost in terms of very poor wages and working conditions.[10]

On the whole, it is probable that the greatest areas of growth in women's manufacturing employment are largely in the unorganized sector in the bigger urban agglomerations, where wages are so low that only women – excluded from the organized sector and desperate for work – are prepared to accept them. Although women do not always lose out entirely in the labour market through technological change – they may obtain niches in certain processes which require 'female' skills – the more common experience is that female jobs have often borne the brunt of mechanization. The tendency in some other industries is to employ women only as a lower-cost substitute for mechanization. This traps them in a depressing cycle of either low-wage employment or its loss through rationalization. When the issue of employment and technology is addressed, it is therefore important that the impact of technological change is disaggregated by gender and that the sex specificity of jobs and tasks is taken fully into account.

Services

The service sector has provided one of the few areas of increased employment for women. But because a large part of this employment is in the unenumerated unorganized sector, the 1971 census data record a very low figure for service-sector employment.

The major areas of growth in the *organized* sector have been in the public sector, where the expansion of education, health and administrative functions has brought into the labour force a new group of women. These are middle- and lower-middle-class college-educated women who have benefited from the increased availability of secondary and higher education since Independence. The largest group of employees are school teachers, followed by medical and clerical personnel. Their choice of employment reflects not only the greater availability of organized service-sector employment (and perhaps a greater commitment by government to increasing women's employment), but also strongly held concepts of suitable work for women from 'respectable' backgrounds. Thus, jobs in which women deal mainly with other women are particularly favoured, or indeed forced upon women. Teaching is the most well regarded of all occupations.

One effect of this preference for 'segregated' jobs has been to keep salary levels for women low. For instance, shortfalls in the public-education system have increasingly been met by the mushrooming of private educational establishments. These employ large numbers of women teachers unable to find jobs in government schools. Working conditions in the less reputable ones are often very poor and rates of pay much lower. A combination of fierce competition for jobs and the 'policing' of their behaviour by families and colleagues provides a strong disincentive to organize.

The *unorganized* service sector is the resort of large numbers of very poor urban women. The majority work in domestic service, which is not necessarily the lowest paid work in this sector. However, long working hours and the highly personalized forms of control involved in the employer-servant relationship make it the last resort for poor women. Although there are no reliable figures, it is thought that there are considerable numbers of women employed in prostitution.

Traditional areas of 'female' service provision, such as marketing and distribution, have declined with the development of modern services. However, a range of market needs created by urbanization – such as the provision and sale of cooked food, liquor and fuel for cooking stoves – is met disproportionately by women. Self-employed females are to be found in a range of poorly remunerated personal services where entry costs are low and where some continuing caste and traditional specializations, such as midwifery and other health-related services, may be found. The very hidden nature of this servicing work renders policy intervention particularly difficult. However, the success of the Self-Employed Women's Association[11] provides important pointers as to how such women can be reached.

WOMEN IN THE INDIAN LABOUR MARKET

Labour-market placement

Banerjee has summarized the results of successive surveys on women in the labour market.[12] She finds that women are heavily concentrated in manual and casual activities and are more prone to unemployment and seasonal variations in work availability than men. A possible exception to the latter was noted earlier in relation to the changing patterns of agricultural labour in some areas, with instances of direct competition between women and men resulting in women gaining employment at the expense of men and a corresponding narrowing of wage differentials. In urban employment, however, women at all levels of skill obtain fewer hours of weekly employment than men and wage/salary differentials are firm.

Wage levels

Women earn lower wage rates than men (at the same levels of qualification) and work fewer hours. This is true even when they are employed in the same industry or operation. Thus, in 1972–3, female earnings averaged 58 per cent of those of men at all skill levels below secondary education. But there were significant differences within skill categories, with illiterate women earning only 48 per cent of the wages of illiterate men, and women with middle-level education and some technical skill improving to 75 per cent of the wages of their male counterparts. Standing's small-scale study showed a similar and even more extreme profile, with women averaging overall 60 per cent of male wages but the poorest paid women earning only 28 per cent of the equivalent male wage; substantial improvements only occurred for professional and managerial employment. Despite formal policies of equality, macro-studies of male and female earnings in the organized sector show consistent differences. This is because of the tendency for women to be situated in the lower echelons of white-collar and professional occupations and to be promoted less.

Gender segregation in the labour market helps to maintain gender-based wage differentials. As Banerjee points out, female-specific tasks are invariably categorized as less 'skilled' and therefore justifying lower wages. Banerjee also points out that enclaves of female employment typically arise in low-capital-intensive operations and that women are frequently employed as a low-cost 'substitute' for capital. For example, Indian pharmaceutical factories subcontract the manufacture of glass ampoules to small-scale units employing a home-based female labour force, instead of making them in a factory using modern machinery.

Education levels

Since Independence, Indian women's access to formal education has increased considerably, both absolutely and in relation to men. In 1981, while urban literacy rates were higher for both sexes, the overall literacy rate for women stood at 25 per cent and was half the rate for men. When the data are disaggregated by class, as well as gender, major differences emerge.

Has the general increase in literacy and access to higher education meant an improvement in women's access to employment? The picture is not encouraging. All-India figures consistently show that female employment rises sharply among graduates and illiterate women. In 1971 there were four times as many male degree holders in formal employment as female and the majority of women graduates were still found clustered in a few 'respectable' occupations. Part of the explanation for this is that although there have been substantial increases in female enrolment in medicine, general science and commerce, the bulk of women graduates still come from the arts and humanities and subjects such as home economics. This may restrict their range of options. There are also social and cultural pressures on young women to use their higher education not for employment but for the marriage market. Women's security has been perceived as bound up with marriage rather than employment. However, there is evidence of an emerging trend in the urban middle and lower middle classes to launch daughters into the labour market because of economic pressures and the need to augment the income of the household. Such pressures may have added to the 'inflation' of educational qualifications in the urban job market as employers increasingly use them as a criterion for even the most modest jobs.

This has led some social scientists to hypothesize that the labour market for women is getting increasingly differentiated for different classes. Poor women find themselves squeezed out of jobs formerly available to them because they lack educational qualifications. However, published empirical evidence of this is still lacking. Besides, the segmented nature of the labour market is unlikely to produce much direct competition between illiterate and literate women. What may be emphasized is the relative disadvantage of both middle-class educated women (in relation to their male counterparts), and working-class women (in relation to both working-class men and middle-class women).

Skill levels

On-the-job training and skill acquisition remain at very low levels for women, which perpetuates female disadvantage despite attempts by a number of government commissions to draw attention to these major factors. In 1971, only 1,200 of the 29,700 technical degree and diploma holders were women. Similarly, figures for 1974 show that across 161 trades, out of 52,500 apprenticeships only 104 had gone to women – and these were in 12 of the 161 trades. Only in very few modern industries, such as electronics, where women are considered to have a 'natural' aptitude for the work, are they likely to get training.

THE IMPACT OF SOCIAL CHARACTERISTICS OF GENDER ON WOMEN'S POSITION IN THE LABOUR MARKET

In the previous sections we have seen that female workers are significantly disadvantaged in the labour market. The question may be asked whether – as some economic models say – this is the result of rational choices made by individuals. For example, do women choose to accept lower wages? Discussions in the previous sections have shown that workers are bearers of social characteristics which play a major role in determining their labour-market position. It would therefore be more realistic to assume that long-term processes structure individual options and that women are constrained by a prior set of economic and social factors.

Women's relationship to the family

Many of these long-term processes which structure the relationship between women and men and the division of labour between them are located within the family. In order to understand women's changing relationship to the labour market, we need to examine their situation in the family. It is equally important to disaggregate women in terms of class, lifestyle and marital status so as to examine how changes in the labour market may affect different groups of women.

Married women's commitments within the family, particularly their overwhelming responsibility for housework and child care, are often seen as primarily responsible for their subordinate position in the labour market. However, it must be pointed out that disadvantages accrue to women as a category regardless of whether they are married or have children. Besides, when analysing this factor, we must be careful not to assume a particular European model of the family. Married women in India may find themselves in different family situations which may vary by class and region. For example, they may be living within extended families with wider networks of kin able to give assistance to the working mother. Or they may be able to call on help from domestic servants. In general, in India, middle-class women are more likely to be able to obtain assistance, unpaid or paid, while working-class women tend to have less access to kin and cannot afford paid help.

Women's commitment to employment

The real and perceived family burdens of married women are often cited by employers as reasons for their perception of female labour as more unstable and less committed than male labour. Women workers are thought to take more time off for domestic duties and to leave the labour force at particular points of time, such as at marriage or childbirth. Employers are therefore less motivated to employ, train or promote women. However, there is little evidence to support these views. Many studies have demonstrated the stability of Indian female labour and have shown that women are less prone to leave their employment for better jobs.

Studies on absenteeism in the organized sector show that rates for males and females are fairly similar. One study found that female absenteeism was usually related to family responsibilities and male absenteeism to alcoholism which is arguably a more serious problem for employers.

Studies on the age profile of the female labour force show that in India (unlike in most developed countries), women's participation rates in the labour force rise steeply up to the age of 30 and then more gradually, peaking at the age of 45. In other words, women's greatest involvement in the labour force coincides with their years of childbearing and rearing. This reinforces the need to improve social infrastructure, such as child care.

Wage discrimination and poverty

Female labour-force participation levels are directly related to the level of household income (except in the case of a small proportion of educated women who seek to have a career). The most impoverished households have the highest numbers of women workers. The acceptance of very low wages by women is due to the fact that women bear the primary responsibility for the welfare of the family. Women in impoverished homes have little choice but to accept the low wages.

In the case of some sections of the middle class, the experience of rapidly declining real incomes has led to the new phenomenon of unmarried daughters entering the labour force. Unlike men, their selection of a particular job is likely to be determined not by the level of remuneration, but by the necessity to augment low family incomes.

Thus for poor women, the need is not just for employment creation but for better paid employment. In this context, the promotion and fostering by successive Indian governments of small-scale unorganized sector industry has increased the number of jobs but not the pay.

Despite the enforced movement of more women into the labour force, gender stereotyping in the labour market is still strong and there has been little change in the attitudes of labour leaders and public and private sector employers. The political climate continues to favour the concept of the male breadwinner and men's greater right to employment. Interventions at the socio-political level are likely to be required to tackle these problems.

POLICY ISSUES

Statistical indicators and the problem of underrenumeration

The discussion of the census data from India points out their limitations in general in providing a reliable account of the sex composition of the labour force and of trends through time. This is particularly problematic where a large amount of economic activity is unorganized. An apparent decline in the officially recorded labour-force participation rate of women may conceal shifts from the organized to the unorganized sector, or independent increases in unrecorded employment. In general, the more aggregated the data source, the greater the unreliability. Sources other than the census, such as the National Sample Survey Organization, which use consistent definitions and smaller sample frames, are a better starting point.

Women and the unorganized sector

Underrenumeration of unorganized sector employment, where women are disproportionately represented, hinders informed policy-making. The lack of information on the unorganized sector particularly constrains the formulation of initiatives which need to take into account the gender dimensions of poverty. Case studies show that female-headed and female-wage-dependent households are concentrated in the lowest income percentiles because of their confinement to casualized, erratic and low-paid work. Women workers are often employed in an isolated way in home-based industry. These women need to be reached and supported with informational and organizational networks to combat their isolation and to provide them with a better knowledge of the market.

Disaggregating women

Policy should not be created around the assumption that women are a homogeneous category. Class, culture and ethnicity, as well as other factors, differentiate women's life chances. It is critical to differentiate processes within the wider economy in order to understand which women are particularly vulnerable to macro-level change. Moreover, the relevance of change has to be understood from the context rather than assumed from first principles.

Labour-market segmentation and wage levels

Both gender-segregated and segmented labour markets have complex effects on levels of employment for both sexes. Women's confinement to primary and low value-added segments may mean that they are protected in some areas of employment, but this is at the cost of low wages and vulnerability to technical substitution. Where women and men are placed in direct competition with each other, women gain employment because they are cheaper. Again, *a priori* assumptions should not be made that an increase in the availability of employment is automatically a good thing. The nature of the employment and particularly the wage levels and work conditions should be examined.

Policy-makers need to address not only the general issue of wage discrimination against women but to disaggregate the causes and address the particularly severe implications for the poorest people.

Skill levels

The disadvantaged labour-market position of women points to the need to intervene at all levels of skill acquisition from the school stage onwards. Positive action to combat the persistent labelling of women as less skilled can be encouraged at the level of the firm. Literacy rates need to be increased, particularly among rural women.

Several attempts and programmes to equip women with skills have tended to reproduce the same stereotype of women as 'naturally' endowed with domestic skills. India abounds with well-intended income-generating schemes which teach women to sew, embroider and make items such as lace tablecloths. Such projects are rarely commercially viable. There have been but few attempts either to utilize or to improve the standard of women's existing skills within the informal economy and to exploit niches in the market.[13]

Social infrastructure

The high level of burden which working women bear for the management of their household duties points to the need for policy initiatives for the improvement of the social infrastructure around women. Issues such as neighbourhood child care provision and safe and efficient transport are often as important for women as workplace issues. Despite a high level of reliance on kin for child care in countries like India, there are substantial numbers of women, particularly among the poor, who lack such support. Many women also face various forms of sexual harassment on the streets and at work, and infrastructural changes are needed to alleviate this problem.

CONCLUSIONS

The Indian case study has shown that gender is an important factor in determining not only labour-market placement and wage levels, but also the ability of workers to train or educate themselves for better jobs.

Jobs and tasks in the labour market are both gender-segregated and gender-segmented. Studies show that women workers earn consistently less than male workers and are found predominantly in manual, low-skilled and casual work. A large part of the female labour force is employed in the informal sector which is beyond the reach of legislation. Governmental policies have helped to increase the number of jobs but not their pay.

Women's commitments within the family are often seen as primarily responsible for their subordinate position in the labour market. However, Indian studies show that women's labour-force participation rates rise up to the age of 30 – that is, during the childbearing and rearing years. It seems more reasonable to assume that women accept low wages because they bear final responsibility for the welfare of their families and need to augment family incomes in order to alleviate the poverty of their household.

It is crucial that development policy-makers and planners realize that the impact of change and policy varies significantly not only by gender, but also according to sector, region, class and other factors. To be adequately informed, planners require gender-disaggregated data backed by micro-level studies with a gender perspective. Without such an input, employment policies will continue to be gender blind.

NOTES

1 Whitehead (1985).
2 Dex (1985).
3 Humphrey (1985).
4 Heyzer (1986).
5 These problems are discussed at length in Chapter 2: *Statistics*.
6 Chatterji (1984).
7 See Billings and Singh (1970).
8 Dyson and Moore (1983).
9 See Baud (1983) for findings on the textile industry.
10 See Banerjee (1985).
11 SEWA (1983) and Bhatt (1976).
12 Banerjee (1985).
13 But see SEWA (1983) and Bhatt (1976).

Chapter 5

Housing

Caroline O. N. Moser

Women are primary users of housing and yet they often participate much less than men in the execution of low-income housing projects. Does current housing policy and provision tend to exclude or constrain their participation? Do women in low-income communities have particular housing problems and needs which are not sufficiently understood by those responsible for housing policies, programmes and projects?

This chapter examines these questions and concludes that current housing policies in Third World countries appear to inhibit women's participation in low-income housing projects. Particularly affected are female-headed households, which are disproportionately represented among the poorest of the poor and account for a sizeable proportion of urban households. Paradoxically, this group is often excluded from low-income projects because of gender-blind housing policies.

In the last twenty years there has been a widespread proliferation of squatter and shanty settlements where the majority of the population of many cities have taken responsibility for their own housing provision. As conventional 'top-down' public housing projects (that is, projects where governmental authorities plan, finance and implement projects without the participation of beneficiaries) have failed to satisfy low-income housing demand in the Third World countries, there has increasingly been a shift in governmental housing policy to a diversity of assisted alternative 'self-help' solutions, of which the most important types are 'sites-and-service' housing projects and 'upgrading-schemes'. Such self-help solutions which involve the participation of beneficiaries in all the aspects of the project are now recognized by international agencies and national governments as offering cheap alternative housing to a large proportion of the urban population without major increases in the proportion of investment allocated to housing. But if such self-help solutions are to be efficiently and effectively implemented and if there is to be equity between men and women in their access to housing, it is crucially important that the participation of women in housing projects is enhanced through gender-aware housing policies.

The first part of this chapter argues that the current constraints to gender-aware housing policies are principally the result of stereotyped assumptions on the part of planners about the structure of the low-income household and the distribution of tasks within it.[1] The second part of the chapter analyses the consequences of gender-blind housing policies. Key issues in gender awareness are highlighted and illustrated with examples drawn from a large number of empirical studies on women and shelter provision. The conclusion shows that present housing policies often fail because they do not take account of women's strategic gender needs in housing.

MISTAKEN ASSUMPTIONS IN CURRENT HOUSING POLICIES

Gender awareness in housing entails recognizing, understanding and identifying the different and particular housing needs of women and men on the basis of gender. Unfortunately, in current housing practice in the Third World, there seems to be a wide tendency on the part of policy-makers, planners, architects and designers to make assumptions which do not fit the reality of women's lives in Third World urban society.

First it is assumed that the household consists of a nuclear family of husband, wife and two to three children. Second, it is assumed that in the family there is a clear division of labour, in which the man of the family, as the breadwinner, is primarily involved in productive work outside the home while the woman, as the housewife, takes overall responsibility for the reproductive and domestic work involved in the organization of the household. Implicit in this is also the assumption that within the household there is equal control over resources and decision-making between men and women in matters affecting the household's livelihood.

The first assumption fails to recognize that low-income households are not homogeneous in terms of family structure. Although nuclear families may be the dominant type, a diversity of other structures may occur. These may vary depending on different stages in individual life cycles and, more importantly, in different economic, geographical and cultural contexts (with considerable variation between Latin America, Africa and Asia). The household structures may include traditional types such as simultaneous and serial polygamy, as well as new types emerging as a consequence of current economic transformations. Of the latter the most common and the most vulnerable are female-headed households, in which the male partner is absent, either temporarily because of migration or permanently because of abandonment, divorce or death. It is estimated that one-third of the world's households are now headed by women. In some urban areas, and especially in Latin America and parts of Africa, the figure reaches 50 per cent or more. What's more, this is an increasing phenomena.

The second assumption fails to recognize that women in low-income households perform a 'triple role'. First, 'women's work' includes *reproductive work*: childbearing and rearing responsibilities. Second, it includes *productive work*, often as secondary income earners, located either in the home (for example in subcontracting or piece-rate work) or in informal sector enterprises, a sector in which the proportion of women is dominant.[2] Third, 'women's work' increasingly includes *community-managing work*, the organizational jobs undertaken by women at the neighbourhood level. As an extension of their domestic work, women are increasingly forced to take the primary responsibility in the organization of basic services at the community level and in the formation of local-level protest groups to put pressure on local government to provide infrastructure.[3]

Understanding this triple workload of women in low-income households is crucially important in developing gender-sensitive housing policies. Unlike men, women are severely constrained by the burden of simultaneously balancing these three roles. Their reproductive and community-managing work is often seen as 'natural' or non-productive and is not valued as work. It is thus invisibilized and it is erroneously assumed that women have 'free time'. Then planners wonder why women fail to participate in self-help housing projects which rely particularly on their contributing work. This is particularly true in the case of female-headed households where the problem of the triple burden can be severe. The lack of gender awareness on the part of the authorities effectively bars these women from participating in self-help projects.

In any discussion of gender, it is also important to recognize that women, like men, are not a homogeneous category. The needs of low-income women differ from other income groups. They also vary according to culture, age, marital status and age in the life cycle.

Therefore women's capacity to participate in housing projects may vary considerably from one specific context to another. Some of the examples given in the next section reflect these differences and indicate the necessity for sensitivity and understanding of the specific context.[4]

CONSEQUENCES OF GENDER-BLIND HOUSING POLICIES AT THE PROJECT LEVEL

Women experience a wide range of problems because of a lack of gender awareness in current housing practice. Here we explore current practices and draw alternative solutions. The analysis pertains to housing provision in both sites-and-services and squatter-upgrading projects. In sites-and-services projects, public authorities intervene in the provision and development of new housing sites with infrastructural facilities, whereas, in the case of squatter-upgrading schemes, intervention is usually limited to improving the infrastructure of existing sites. The purpose of the analysis is to provide tools to translate gender awareness into practice. Highly complex issues are simplified to illustrate the way in which development practitioners may themselves intervene with greater awareness of gender. Nine phases at the project level are selected for discussion. These phases can be divided into three groups, representing the three stages of project formulation, implementation and management. If projects are to succeed, women's participation in these phases is considered particularly important.

Project formulation

Formulating eligibility criteria

Formulating eligibility criteria is one of the first and most critical stages in the execution of low-income housing projects, as it determines who is going to benefit from the project. This stage is particularly relevant to sites-and-services schemes rather than to squatter or slum-upgrading programmes.

Income is one of the key criteria for selection, as the aim is to provide shelter for those who are too poor to afford 'conventional' housing, but not too poor to make regular repayments for land, housing and services. Minimum earned-income levels are specified and in general the poorest 10 per cent of the population is excluded. Female-headed households are disproportionately among the poorest of the poor and are frequently excluded. For example, in Quito, Ecuador, 30 per cent of the total applicants in the Solanda sites-and-services project were female-headed households.[5] However, because of the level of monthly earnings (between 7,890 sucres/$131.5 and 10,716 sucres/$178.6 at the 1983 rate of exchange) and the initial down-payment stipulated by the eligibility criteria (which represented 5–15 per cent of total housing costs), nearly half (46 per cent) of the female-headed households did not qualify for access to the scheme.

Income criteria based on earnings alone ignores transfer income (i.e. economic support from friends and relatives, loans of food and clothing, etc.) which often make up an appreciable part of the real income of women-headed households. It also ignores considerations of willingness to pay for housing, which recent studies[6] have shown to have a definite gender bias. Women have a greater need to secure their dwelling and take a more responsible attitude to housing loans.[7]

Secondary criteria for defining project beneficiaries, like *proof of fixed employment* and *stipulations about family structure*, can further discriminate against female head of households who are predominantly in informal-sector jobs.[8] In a Brazilian project (Boa Vista in Vila Velha) it was stipulated that applicants should be fathers of at least two children, thus assuming that heads of household would be male and that households would conform to the Western nuclear model.

Given the economic situation of female-headed households and their increasing numbers in most Third World cities, this group can hardly be ignored as legitimate beneficiaries of housing policies. As the present eligibility criteria take no account of the constraints placed on their productive roles by the demands of their other roles or their difficulties in the labour market, other alternatives will have to be sought. For example, a certain number of plots within a site-and-service programme could be reserved for female-headed households at a reduced cost. Alternatively, the upgrading of their present shelter has been found to be an appropriate solution.

Recruitment of beneficiaries

Once the eligibility criteria are fixed, the beneficiaries of the project have to be recruited. Here again, discrimination occurs. Two aspects of recruitment are discussed: the means of advertising the project and the application procedures.

In *advertising* projects, public authorities often distribute information in the written form, i.e. in newspapers or public notices. As women in the Third World have considerably less access to education than men and are geographically tied to the home or the neighbourhood, this kind of information is less likely to reach them. This is particularly true if the notices are made available through the workplaces or public meetings. For example, in a study of three housing projects in Paraguay, nearly half (44 per cent) of the women interviewed in the three communities complained that their general lack of information about shelter was a major obstacle to improving their living and housing conditions, whereas only 16 per cent of men identified this as a problem.[9] Even where more direct methods of advertising are used, such as local meetings in the community where eligible applicants are likely to live, women may be prevented from attending by constraints of time (due to their triple role) and by cultural factors which inhibit public participation.

Alternative forms of communication could include verbal information through community level organizations (for example church groups and child-care clinics) and through informal gatherings where women are likely to congregate (such as markets or water points).

Once people have learnt of the scheme, *application procedures* have to be completed. Filling in the forms and providing supporting documents can be particularly problematic for women. For example, in the Solanda project in Quito, of the 7,176 women applicants who responded to a socio-economic survey, 33 per cent of the original applicants dropped out even before the selection had begun because they were unable to complete documentation requirements.[10]

To assist women in managing the logistics of completing and substantiating application forms, project staff should attempt to reduce women's fears of formal institutions, minimize and simplify paperwork and personally assist in questionnaire completion. Bureaucratic activities can be decentralized to areas where the target group lives. Other helpful actions are the provision where necessary of house-to-house visits, child-care facilities and the use of women professionals.

Planning the settlement and the infrastructure

Issues that arise from planning the settlement affect all women regardless of their marital status. As women are the prime users of housing and settlement-level infrastructure, settlement planning affects them more than any other group and their inclusion in decision-making at this stage is critical. Three aspects of settlement planning which particularly affect women are location of the housing site, tenure/ownership of the dwelling and the settlement-level infrastructure.

Decisions about the *location* of low-income housing sites are generally made by project authorities. Trade-offs are made between costs of land, transportation and links to existing infrastructure. A critical element is the work location of the target group. Despite the importance of women's earnings, the location of women's work is rarely considered. In trying to combine economic and domestic work, women have little time to travel and their ability and choice of work is particularly sensitive to housing location.

Where settlements are forced to locate in areas that discriminate against women's employment patterns, provision should be made for cheap, reliable and frequent transport, and/or for the establishment of income-generating opportunities for women, as well as schools, markets and child-care facilities within the settlements.[11]

Tenure is a critical factor in both sites-and-services and upgrading projects. Rights to land are usually given to men on the assumption that they head households. However, there may be at least three reasons why ownership should be extended to women. First, since women have the ultimate rearing and caring responsibilities for their children, in cultures where there is a high incidence of marital separation, rights to land can act as a way to safeguard the families' interests. Second, ownership can strengthen the position of women within the family unit and this in turn would increase their participation in decisions and activities relating to the housing project. Third, land titles can enhance women's access to credit which is usually very limited.

Insufficient attention to women's culturally specific economic and domestic needs in the development of the *settlement-level plans and infrastructure* can result in a grave deterioration of their lifestyles and living conditions. Planners all too often ignore the needs of users and tend to emphasize technical, financial or economic criteria for the most efficient use of space.

There are many case studies to illustrate this point. For example, in the relocation area of George, Lusaka, roads and plots were laid out in a grid-iron pattern with each house on a separate plot. This isolated the women from each other and deprived them of the more sociable and mutual help environment in which they were used to performing their domestic chores.[12] In another case, blanket residential zoning instituted by the authorities in the Dandora sites-and-services project jointly funded by the Kenyan government and World Bank had severe implications for many women who were informal-sector workers operating from home. Despite threats and harassment from the City

Commission, 48 per cent of these women continued to operate small enterprises such as selling vegetables and charcoal from their houses.

In the provision of services at settlement level, women as the main users of them may be in the best position to decide the relative merits of different types of services. Women may give priority to water supply and sanitation services over electricity provision in order to reduce their workload significantly. When new services are introduced without consultations with the users, they may prove inappropriate or be underused. In many countries clean water pumps placed in low-income communities were not designed for women and children, their main users, and have consequently broken down.[13] Toilets have been underused by women because of a gap at the bottom of the wall which exposed their feet and offended notions of privacy.[14]

Women's participation in decisions regarding services and the location of settlements is limited by the fact that such decisions are often made before the beneficiaries are selected. Another difficulty is that a community has to be created among people who probably do not know one another. A survey of a suitable proxy group for the target group could yield valuable results about women's needs and priorities. If project staff are gender-aware, they can develop organizations appropriate to the participation of women from the beginning.

Planning the dwelling

Women are significantly affected by the way the dwelling is planned, yet their control over decision-making in this area may be restricted. In male-dominated households women may be excluded from decisions although their spouses know little about housework and spend little time in the dwelling. Community-level politics may also exert influence on the nature and extent of women's participation. Most importantly, planning authorities often dictate house design, standards and guidelines within which beneficiaries may build.

Guidelines set by public authorities include *minimum housing standards* and *time limits* for completion. These exclude the possibility of using indigenous and recycled materials and make the construction costly and time-consuming. Female-headed households are particularly disadvantaged by the added costs and deadlines. Moreover, they may not have the required skills nor the ability to hire skilled labour to build to specified standards on a self-help basis.[15] There may be a case for identifying sub-groups within the target population on the basis of the level and regularity of income, number of dependants, etc., so that standards and time limits can be negotiated according to the capabilities of each household.

When guidelines try to replace traditional styles of dwelling they may force changes in lifestyles. This happened in a 'Better Housing Campaign' in Tanzania, which made it difficult for family members to live separately in accordance with their traditional concepts of privacy and autonomy. Women, in losing their traditional autonomy, were redefined as dependants.

The design process is a major area for female participation in project formulation. In specifying model *house designs*, public authorities often do not appreciate the ways in which culturally insensitive designs can have negative effects on women and their families. In a Tunisian government/USAID project in two settlements in Tunis, women were dissatisfied with the small size of the inner courtyard. As female social life was almost entirely confined to the home, the reduced inner courtyards led in some cases to psychological depression, neuroses and even suicide among women.[16] On the other hand, women in a Nicaraguan project in the San Judas expressed a preference for more space outside. Thus only 20 per cent of the plot was built on, leaving the rest of the land free for small-scale cultivation, traditionally women's work and an important component in family subsistence.[17] Another important factor in house design that planners frequently fail to recognize is the need for women to generate income in the home, either in the form of renting out space or from informal enterprises.

Financing housing

House-building in both site-and-services and upgrading projects is generally self-financed by residents. The levels of investment vary according to the labour provided by the beneficiaries, the amount of hired labour and the type of building materials designated as appropriate by project authorities.

While finance affects all applicants, female-headed households are often hardest hit. In fact, financial constraints may be the prime reason for their exclusion or their high drop-out rate. Five principal constraints are discussed below.

First, *savings* are required for downpayments, alternative housing during building, building materials and hired labour. Female-headed households may not be able to meet all of these demands.

Second, in many countries, *restricted access to credit* constrains women in financing housing. They face obstacles such as lack of information, low and irregular incomes, lack of collateral, inability to deal with complicated loan procedures, discrimination by male bureaucrats, high interest rates and women's lack of legal standing. Ways of overcoming some of the disadvantages could be land ownership for women (as discussed earlier), informal credit institutions for women (e.g. SEWA in India) or a project-loan fund with procedures and rules established in accordance with the particular needs of different household types.

Third, in many projects the responsibility for *purchasing building materials* is on the individual households. Female-headed households lack both the time to shop around and the money to pay high delivery charges. To overcome such problems a building-materials centre with competitive prices and perhaps free delivery and interest-free credit could be set up by the project authorities.

Fourth, *loss of earnings during construction* can be a severe constraint to participation in housing projects. In an attempt to overcome this, the women's Self-Help Construction Project in Panama, launched in 1981, provided a number of supports for its participants, 45 per cent of whom were female-headed households. These included subsidized building materials and tools and a stipend for women during the construction phase, financed by a loan to the women with a 25-year repayment period to the women.[18]

Fifth, where *allocation of household budgets* is controlled by the male head of household, women may suffer if the household finance is unduly cut back to allow funds for the housing project. Some form of subsidizing of basic needs – such as food, health care and child care – has proved useful. This technique was used in the San Judas project in Nicaragua.[19]

Project implementation

House construction

Women participate in the actual construction of the dwelling. The nature and extent of their participation depends on the household structure and the type of self-help construction. Three types of construction projects are discussed below.

Projects based on *individual self-help* usually assume that the majority of the urban poor are unemployed or underemployed and have lots of time. However, not all households have equal time or skills.

Female-headed households are often severely stressed. Lacking time and skills, as discussed earlier, they may be forced to hire professional labour they can ill afford in order to meet the minimum standards and deadlines set by project authorities. Full participation of women is often constrained by the attitudes of men against their participation in 'men's work', and the women end up in a 'secondary' role, doing such heavy tasks as carrying materials and water, preparing cement and digging, etc.

In the case of projects based on *mutual self-help*, construction materials can be bought in bulk, community skills can be pooled and the project staff can intervene to a greater extent to involve women more fully. For example, in the Panama Women's Self Help Construction Project, 99 of the 105 women in the project received training from the National Service for Professional Training (SEAFORP) to build 100 houses in Curundu. Work groups consisted of eight to ten women supervised by four specialists. Construction was completed successfully and houses were allocated by mutual agreement to those who were in greatest need and had worked the hardest.[20]

However, not all group work is so harmonious and gender conflicts often arise. The case study discussed below, of the San Judas upgrading project in Managua, Nicaragua, shows how conflicts arose because of the attitudes of the team leaders (all men) and the lack of skills and confidence of the women.

Case study: women's participation in self-help housing

This case study by Irene Vance assesses the extent to which conflicts arose between men and women in the San Judas Barrio Project, how they were solved and how they modified the original proposals.

The settlement of San Judas is one of the oldest and most densely populated 'barrios' in Managua. The San Judas barrio committee (CBS), which consisted of an equal number of men and women, was actively organizing the community-salvage operation from the war damage after the Sandinista victory in 1979. Maria Silva, a representative on the barrio committee, was elected to take special responsibility for housing and community development. Through her efforts a housing donation of one million cordobas was acquired from an international religious organization. The donation was to be channelled through the Ministry of Housing and Human Settlements (MINVAH).

The criteria for choosing beneficiaries were overwhelmingly based on need and a selection of 60 families was made by the barrio committee. Of the 60 families, almost half were female-headed households, 22 of whom were single women supporting their children.

Discussions finalizing the design features of the project were taken in a series of collective meetings in which the architect presented several alternative models. These meetings were mostly attended by women. The majority attitude among the men was that 'all you get is a lot of chat, a lot of women talking. We can attend when the real work begins'.

Before the construction-work building commenced, the weekly meetings were given over to a training programme. Several sessions were devoted to discussing issues related to cooperation, organization of work-teams and collective responsibility. Again the majority of those who attended were women. Many men stayed away, considering that 'this wasn't real work, just more chat'.

According to the Ministry, both men and women were to share equally the various tasks of construction and work-teams were to be selected taking into account such factors as level of political awareness, sex, economic activity and previous experience of construction work. Four team leaders, all of them men, were chosen on the basis that they had building experience and could train the others. One of the team leaders decided that each team would consist of male officials and women working as helpers (in marked contrast to the policy of equality that the MINVAH had planned).

The first stage of implementation was to prepare the worksite. There were no written rules and only an understanding that as many beneficiaries as possible should turn up for the very tiring work. Records show that women's attendance was consistently higher than men's.

Within a month of the start of the building work, the monitoring reports of the ministry show that work was not functioning properly. No training was taking place, there was uneven attendance, there was poor performance on the part of women and too many women were on the building site. The team leaders recommended that women should send men to represent them. This provoked an angry response from the women who found it biased and complained that the team leaders were at fault for not taking their training responsibilities seriously.

The draft contract of work was reformulated by the ministry, giving more influence to men on decision-making and reaffirming the proposal that women send men to represent them on the site. Many women among the beneficiaries had no available men, nor could they pay for men to represent them and they resisted signing the contract. Moreover, several men had to leave because of illness or military service. It soon became difficult to maintain a regular workforce.

Gradually the women gained recognition for their commitment and good work. They also acquired skills, confidence and the respect of the men, which improved relations between the team leaders and the female workforce.

There were gender-related discussions on the question of the allocation of dwellings, with women wanting the allocation on the basis of need and the men on the basis of time and work invested. The ministry obliged the collective to find its own solutions to disputes. Work continued despite the disputes and the project was completed successfully.

As can be seen from this case study, an important element in women's participation in self-help housing projects is their need for training in building skills. In a Save the Children Fund (SAVE) squatter-upgrading project started in Kirillapore, Colombo, Sri Lanka, in 1979, accent was laid on training women in masonry and building skills. As a result, women headed three of the five work teams and accounted for 38 of the 55 masons.

Where individual self-build or mutual self-help cannot meet all construction needs, residents may have to hire professional labour, particularly for specialized tasks. In the case of such *contract-built housing*, project staff should be ready to give technical support to assist residents in the hiring of labour, in the supervision of work and in the assessment of completed building.

Obtaining community-level services

Obtaining services like water, sanitation, electricity, roads and social services from local authorities requires a considerable amount of effort and organization at the level of the community. This is particularly so in the case of squatter and slum-upgrading projects. Sites-and-services projects often include such community-level services in the overall package.

Women, because of their reproductive activities, are affected most by the lack of these services. They are therefore the most committed to obtaining these services. Thus in spite of the burden of their triple role, the women are frequently expected by the community and the planners to take the prime responsibility in this area. In doing so they often jeopardize their other work. Here again women lose out because of planners' false assumption that they have 'free time'.

Mobilizing community action and lobbying local authorities can be very time-consuming and involve years of work. The following quotation from a case study undertaken by Moser in 1978 on the participation of women in the Indio Guayas barrio-level committee in the mangrove swamp periphery of the city of Guayaquil, Ecuador, shows what a heavy added burden women are expected to bear.

> The most important 'external' function of the barrio committee is to petition for infrastructure. Forms of mobilisation are determined by the way in which the Municipality allocates its limited resources on the basis of patron–client relationships. This requires local committees to be co-opted by political parties which exchange services in return for votes and political support. The different stages of mobilisation involve considerable time consuming 'voluntary' work. During the co-option process there are a succession of lengthy meetings within the barrio, as well as with party representatives and local government officials. The preparation and presentation of complex petition documentation is undertaken as well as the very rapid organization of 'spontaneous' large scale mobilisation to protest at the Mayor's office in City Hall at propitious moments. Since the provision of infrastructure is ultimately in direct exchange for votes, the barrio committee has particular responsibilities to the political parties at election time. These include attendance at party briefings, organization of bus loads of supporters for political gatherings and extensive barrio level canvassing. Finally, when infrastructure is provided, the committee must ensure that the community's plan of work is implemented and that the neighbouring committees do not manage to divert the infrastructure through bribes to implementing agency officials. In local level committees it is the women, both presidents and rank and file members, who bear the primary responsibility for this work.

Project management

Project maintenance

Maintenance of housing and services is crucially important in reducing additional expenditure and in increasing self-reliance among residents. Women, as an extension of their domestic duties, bear the main burden of maintenance. It is therefore very important that planners recognize this. Projects where women are not involved at the various stages of planning and construction often show poor results in terms of deterioration of the settlement.

When new infrastructural technology is introduced, it should be appropriately designed for use and maintenance to ensure that facilities do not break down and are not underused. In 'top-down' projects in particular, project staff do not always recognize the work women do in maintaining the community.

Cost recovery

Housing projects of both the sites-and-services and the upgrading types commit residents to repayment of loans for land, house-building and/or services over a period of 10 to 30 years. Cost recovery can affect women in different ways.

Where *ability* to pay is calculated on the basis of average household earnings, it can put women at a disadvantage because of their insecure informal sector jobs and their low remuneration. A number of schemes adjusting interest rates and pay-back periods to support women have been developed. These include negative amortization techniques,[21] revolving fund/deferred payment options, periods of non-payment at the start of the project,[22] and the use of labour contributions by women for the installation and maintenance of community services. Alternative strategies include the promotion of income-generating activities and the encouragement of informal savings networks.

Studies show that *willingness to pay* is a vital ingredient for successful cost recovery and that women are generally more willing to pay and are more responsible about their debts. Female members of male-dominated households sometimes make personal payments to ensure the shelter for their children and themselves. Women's willingness to pay also appears to be related to their satisfaction with the project. In a study of an Indian credit programme geared to women, it was found that female participation in planning and implementation greatly increased the pay-back rates on loans.[23]

CONCLUSIONS

The analysis of the consequences of gender-blind policies at the project level clearly show that it is not sufficient to recognize women as main users and maintainers of housing at the household and the community level. It is essential that planners also appreciate the triple role that women are balancing. Otherwise, projects that may genuinely want to involve women may have poor and even disastrous results.

It is also essential to appreciate that women in differently structured households cope with their triple role under different conditions. Moreover, women may behave differently in different geographical and cultural contexts.

Women with partners are often constrained by the attitudes of their partners. Planners, project staff and male-dominated community organizations tend to reinforce these attitudes further. Hence women get 'locked in' to traditional roles. Thus the potential resource base of the housing project is not optimized, nor is the satisfaction of the participants.

The economic constraints of housing projects on households with women as primary and often sole earners are particularly severe. Such households are either excluded from the start or can be so overburdened that in the absence of any special support, they may have to drop out of the project. Since this group is a primary target of housing and development policies, far more care should be taken to ensure their participation. Useful strategies, backed by experience and evaluation studies, are available.

Practical and strategic gender needs

In developing gender-aware strategies, it is useful to distinguish between the practical and the strategic gender needs of women.

Women's practical gender needs are those which arise by virtue of their gender in the existing division of labour. For example, adequate housing or a supply of clean water are practical gender needs of low-income women. Although such needs affect all household members, they are identified as 'women's needs' because they reinforce their responsibility for reproductive work in the household.

Women's strategic gender needs are those which arise out of their subordination and are formulated in terms of a more satisfactory organization of society. For example, the removal of institutional forms of discrimination to give women equal rights to land ownership and access to credit are identified as strategic gender needs of women. Since such action challenges the existing division of labour, strategic gender needs are generally considered 'feminist'.[24]

This distinction between two sets of legitimate gender needs is useful in defining the limitations and the potential of housing projects and policies. For example, the provision of housing meets the practical gender needs of women. However, where tenure of housing is given to men on the assumption that they are the heads of household, housing provision will not meet women's strategic gender needs. [25]

This distinction can also be used to clarify the different policy approaches to women in current housing policy.[26] The first is the *welfare approach*, which focuses on enhancing women's reproductive role, seeing them as passive beneficiaries of development. The second is the *equity approach* which focuses on reducing the inequality between men and women through 'top-down' government intervention. The third is the *anti-poverty approach*, which focuses on the productive role and tends to isolate poor women as a category. The fourth is the *efficiency approach*, which is concerned with focusing on the delivery capacity of women to make projects more efficient and effective. The fifth can be termed the *empowerment approach*, which focuses on reducing inequality between men and women through 'bottom-up' women's mobilization.[27]

Only the equity and empowerment approaches recognize women's triple role and directly challenge the traditional gender division of labour. In this sense they are aimed at meeting women's strategic gender needs, while all the other approaches aim at meeting the practical gender needs.

Current housing policy reflects a mixture of the three approaches which aim at meeting the practical gender needs. The welfare and the anti-poverty approaches are most common. The efficiency approach became increasingly popular in the 1980s and was actively promoted by agencies like the World Bank.[28] However, without recognition of women's triple role and hence their strategic gender needs, women's participation in housing projects will remain constrained.

NOTES

1 Moser (1989).
2 See Young and Moser (1981).
3 See Moser (1989).
4 For example, see pp. 82-3.
5 Blayney and Lycette (1983).
6 See Chant (1987).
7 See pp. 84-5.
8 See Machado (1987).
9 Sorock et al. (1984).
10 Lycette and Jaramillo (1984).
11 UNCHS (1984).
12 Schlyter (1984).
13 IWTC (1982).
14 IWTC (1982).
15 UNCHS (1984).

16 Resources for Action (1982a).
17 Vance (1987).
18 Girling et al. (1983).
19 See Vance (1987).
20 Girling et al. (1983).
21 Girling et al. (1983).
22 Vance (1987).
23 Singh (1980).
24 Moser (1989).
25 Moser (1989).
26 Moser (1989).
27 See Moser (1989).
28 Moser (1989a).

Chapter 6

Transport

Caren Levy

THE IMPORTANCE OF GENDER IN PLANNING URBAN TRANSPORT

The role of investment in urban transport infrastructure has long been recognized by national governments and international agencies as an important instrument of development. However, transport planners and engineers have traditionally focused on meeting goals related to the operational efficiency of the transport system itself rather than on goals related to development. This is based on a view of their discipline as 'technical', and '... the belief that somehow, in the improved operational efficiency of transport systems (in terms of increased speed, reduced travel time etc.), there will be an automatic, positive developmental spin-off effect'.[1]

This approach led transport planners and engineers to treat transport models developed in North America and Europe as universally applicable in the urban centres of Asia, Africa and Latin America. In addition, these models were applied without consideration of their economic, social or environmental impacts. Therefore, from a development point of view, traditional transport planning has not taken into account the needs of different income and social groups. As a result, low-income households are highly disadvantaged and low-income women in particular are often excluded from transport provision in urban centres.

In the provision of urban transport, many governments and international agencies have recognized the need for planners to differentiate target groups on the basis of income. Low-income households have different transport needs from other income groups. With increasing urban sprawl they are generally located on settlements and sites on the periphery of urban centres. Therefore their capacity to bear the additional burden of time, effort and cost in transport is lower.

In recent years governments and international agencies have also recognized the need for planners to take environmental concerns more seriously in the provision of urban transport. Increasing evidence is emerging not only of the environmental damage of particular transport modes and their related infrastructure, but also of their negative impact on the health of particular groups in the urban population.

It is also becoming evident that planners need to further disaggregate low-income communities and to distinguish target groups on the basis of gender: that is, in terms of the different needs of men and women. Women comprise half the population of low-income communities and are extensive users of basic services such as transport. Men and women, by virtue of the different roles they play in society, have different needs in urban transport.

In examining the implications of gender needs in transport, it is important to remember that women are not a homogeneous category. Not only do the needs and priorities of low-income women differ from those of other income levels, but they also vary according to their stage in the life-cycle and their culture. For example, married women with children have different needs with respect to the use of transport than single women or women who head households. Moreover, different cultural factors affect the pattern of women's activities and, thus, the extent and destination of their travel outside the home. These variables highlight the need to understand the context in order to plan transport effectively. These factors challenge the universal and often implicit assumptions made by planners in their engineering-transport models, particularly in relation to the gender needs of low-income households.

GENDER ASSUMPTIONS UNDERLYING THE PROVISION OF URBAN TRANSPORT

Policy-makers, planners and engineers often fail to recognize the different and particular transport needs of women and men because they make implicit assumptions about the structure of the low-income families, the division of labour within the family and the control of resources and decision-making.[2] First, they assume that the household consists of a nuclear family of husband, wife and two to three children. Second, they also assume that in the family, the man is the 'breadwinner' and the woman is the housewife or 'homemaker'. In these assumed roles, the man is primarily involved in productive work outside the home, either in a factory or in the informal sector, while the woman takes overall responsibilty for the reproductive and domestic work involved in the organization of the household. Third, they also assume that the man and the woman have equal control over resources and equal power when it comes to making decisions in matters that affect household livelihood and well-being. For a fuller discussion of these points see Chapter 5: *Housing*.

Each of these stereotyped assumptions can be challenged by the empirical reality of women's lives in the Third World.[3] They fail to recognize that low-income households are not homogeneous in terms of family structure. Although nuclear families may be the dominant type, there is in fact a diversity of structures based on life-cycle stages and, more importantly, on different economic, geographical and cultural factors, with considerable variation between Latin America, Africa and Asia. These structures may include more traditional ones like simultaneous and serial polygamy, or new ones which are becoming widespread as a consequence of rapid economic transformation.

Among the latter, the most vulnerable are female-headed households in which the male partner is absent either temporarily (because of migration for work or as a refugee) or permanently (because of abandonment, divorce or death). It is estimated that one-third of the world's households are now headed by women, but in some urban areas of Latin America and parts of Africa the figure reaches 50 per cent or more. What is more, this is an increasing rather than a declining phenomenon. As female-headed households are found to be disproportionately represented among the poorest sections of society, transport policies and programmes which do not take account of the needs of women have a negative impact on those extremely vulnerable families.[4]

The planners' stereotyped model also fails to recognize that in most low-income households, women's 'work' includes:

- Reproductive work: childbearing and rearing responsibilities, and organisation of the household
- Productive work: often as second income-earners (for example, subcontracting or piece-rate work can be located either in the home or in informal enterprises)
- Community-management work: jobs undertaken by women as an extension of their reproductive roles at neighbourhood level to organize and provide basic services and to pressure local government to provide infrastructure such as roads and water.[5]

In failing to identify these roles, planning models do not recognize that women carry the burden of simultaneously balancing their roles.

Moreover, in the case of female-headed households, the burden of the triple role is exacerbated. Both women and men require transport in order to perform many of the activities related to their roles. The failure of transport planners and engineers to recognize the roles of women and men has important implications – as Part III will illustrate – for the way in which the users of transport are identified, the spatial and temporal characteristics of trips are defined, and particular transport modes are prioritized.

The third problem with traditional planning is the treatment of the household as a single unit, in which there is a common set of interests derived from equal power in decision-making. This is reflected in conventional transport planning where travel information is only collected from the head of household, as representive of the household travel characteristics.[6] In fact, different household members – men, women, young children or the elderly – have different travel needs, all of which are important in contributing to household welfare.

Moreover, empirical studies have shown that the household cannot be treated as a unit in which there is equal control of resources. In the case of transport, for example, studies have shown that in households where there is a private car, men usually get priority for its use.[7] It is important, therefore, to disaggregate the household so that different needs, priorities and influence in decision-making of different household members can be identified and targeted.

GENDER ISSUES IN URBAN-TRANSPORT PLANNING

With a few exceptions[8] much of the research and planning focusing on gender and transport is confined to North America and Europe.[9] Nevertheless, because transport planning in developing countries has often been practised by international consultants or local planners trained abroad, as though it were universally applicable, urban transport plans often exhibit the same features. However, the impact of these plans differs according to the particular social, economic and political conditions in each city. A comparison of the impact of transport provision on low-income women and men in different parts of the world can therefore tell us a great deal about the requirements that a gender-aware urban-transport planning process would have to meet.

Six different issues in the urban-transport planning process are considered. These are: formulation of goals and objectives, identification of travel patterns, cost of public transport, conditions of travel, community involvement and employment in the transport sector.

Formulation of goals and objectives

Three common characteristics underlie the goals and objectives of most conventional models for urban-transport planning. These are the focus on mobility, on the private car and on the journey to work. The bias with respect to income and gender is clearly apparent.

Although a shift is gradually appearing, most urban-transport plans stress mobility rather than accessibility. The distinction between the two is crucial, 'since the real test of the (transport) system is whether people can obtain access to their activities without undue expenditure of time, money and effort. Accessibility, not mobility is what matters.'[10] Accessibility is based on the relation between the transport system and the pattern of land use, i.e. the spatial distribution of activities such as work, home, shopping, health, etc., which are the many elements of women's triple role. A recognition of this is crucial, because the spatial separation of activities is increasing in urban areas. This is leading to a greater expenditure of time, money and effort on transport by men and women and is resulting in a new set of demands which the existing infrastructure cannot meet. The increasing urban sprawl and specialization of land use is causing enormous difficulties in providing transport, especially for low-income households.

In developing countries, low-income areas are particularly affected, as they are on the periphery of cities where transport is poorly provided. For example, in four Brazilian cities (Rio de Janeiro, Porto Alegre, Belo Horizonte and São Paulo), the average earnings in the peripheral areas are 30 per cent to 50 per cent lower than in the centre. These poor areas are badly linked to the main urban-transport infrastructure, resulting in constrained access to city-wide activities which are important in the satisfaction of the needs of their inhabitants.[11]

In this context, women, and low-income women in particular, are disadvantaged by the mismatch between land-use patterns and transport provision. As will be illustrated in the following sections, there are marked differences between the amounts of 'time, money and effort' that low-income women and men spend on travel. However, whether or not transport planners consider the relationship between land use and transport, stereotypes about gender go largely unquestioned. Mobility or accessibility are still perceived primarily in terms of the male breadwinner – and more particularly of getting him to work by car.

Urban planners' emphasis on mobility is related to their emphasis on the use of the private car. Provision for the private car has dominated planning without regard for the important spatial, social, economic and environmental consequences. This is so even in developing countries, at the expense of cheaper public transport and in spite of the obvious income and gender bias.

The focus on the journey to work in urban-transport planning is unquestionably that of the male breadwinner, based on the stereotype of the nuclear family. It is only since married women have entered the labour force in large numbers that the debate about the differences between men and women's travel needs in the USA, UK and Europe has been stimulated:[12]

> US planners . . . have routinely assumed that new workers entering the paid labour force will behave much the same as historical members of the work force and that their behaviour can be predicted by the usual social indicators of education, occupational status, etc, and not at all by gender.[13]

Another key assumption that underpins transport planning is that the household will locate itself near the man's place of employment. However, empirical data on low-income households in developing countries shows that it is the cost of housing rather than the cost of transport which determines their location, for example on the periphery of urban areas.[14] Transport planning has not responded to this reality. As long as the focus of transport planning remains on the male journey to work in peak hours, the difficulties of access to work and non-work activities for low-income women will continue to be compounded.

Identification of travel patterns

In the identification of travel patterns, techniques based on a 'standard package' have tended to dominate the planning process. Using standard package techniques, forecasting models deal with the generation and distribution of trips, the modal split (i.e. the split between different modes of travel) and the assignment of routes. The collection of data on travel characteristics provides the basis for these models.

Collection of data

Data used in the forecasting models is based on treating the household as a unit and is usually collected according to the stereotypes and assumptions mentioned earlier. Household characteristics normally examined are size, occupation, automobile ownership and income of the (male) head.[15] Little information is gathered about women's access to work, or about the non-work, off-peak and non-vehicular trips undertaken mostly by women for the benefit of the household.

Trip generation and distribution

Studies in urban areas indicate that low-income women make fewer trips than men, that the purpose of their trips differ and that the spatial and temporal pattern is different.[16] These differences are a function of the needs and constraints stemming from the triple role, the transport system itself which is not responsive to women's needs, and the spatial implications of the gender division of labour in employment.

The journey to work accounts for the largest proportion of trips carried out by men and women in low-income households. As fewer women work outside the home than men, their share in the total of work-related journeys is smaller. Moreover, there are important differences between men's and women's journeys to work because of their concentration in different types of work. A high proportion of low-income women work as domestic servants, seamstresses, laundresses, sales clerks or in informal-sector activities located in commercial centres or higher-income areas. For example, a survey of a low-income residential neighbourhood situated near an industrial area in Belo Horizonte, Brazil, showed that 40 per cent of the men as against only 17 per cent of the women were employed in the industrial area. Nearly half the women in employment worked on jobs some distance away from where they lived.[17]

In the case of non-employment-related trips, women and men account for an equal share. The Belo Horizonte study showed that for women these trips were important to the functioning of the household. When women were asked about the purpose of their last trip, 41 per cent were found to be related to health, 19 per cent to household provisioning and 19 per cent to leisure.[18] A key factor in non-employment-related trips is that they are undertaken in off-peak periods. As the concentration of transport provision is for journeys to work in the peak hour, people experience long delays at off-peak hours. This further overburdens women in the performance of their triple role. In particular, extra travelling time can constitute a particularly severe burden for female heads of households whose ability to combine their productive and reproductive roles may be jeopardized.

Levels of service in peak and off-peak periods are calculated on the basis of economic cost/benefit principles. These calculations do not include the value of women's domestic-related trips in off-peak periods, which are often made for the benefit of the household. A typical underlying assumption is that women have 'free time', so that the time factor is seen as less critical in their off-peak journeys. Giving the time women spend in performing their roles some value in the economic costing which transport planners use to guide their decisions on the levels of service is a complex but important part of moving to more gender-aware transport planning.

Women tend to undertake more multipurpose trips than men, combining different household errands where possible. Studies in the USA and Sweden indicate that women working full-time make fewer multipurpose trips than those working part-time and the latter make fewer multipurpose trips than those not employed.[19] Low-income women in developing countries may have similar patterns.

These studies show that women's travel patterns are different from those of men. Thus it is critical that planners distinguish between men and women when focusing on the accessibility of low-income households to urban activities.

The relationship between transport and land use is also critical to more gender-aware planning. In the long run, more gender-aware land-use (or spatial) strategies could reduce the number and length of women's trips and increase their choice of jobs and services. In the short run, temporal strategies like more flexible working hours, longer opening hours (for services), etc., can have a direct impact on releasing time for women to cope with their triple role.

Modal split

Urban-transport planning has focused on the use of the private car and in developing countries this has been done at the expense of cheaper public transport. This strategy has not only an income but also a gender bias. Low-income households are less likely to have cars than high-income households and low-income women are less likely than men to have the use of the car. A study of the transport mode used by the head of the household in Nairobi, Kenya, revealed that while 24 per cent of male heads of household used the private car, only 9 per cent of women heads did.[20] Similar proportions were found for the journey to work in Belo Horizonte, where 23 per cent of trips made by men and only 6 per cent made by women were by car.[21]

The gender bias occurs because male breadwinners usually get priority in using the family car,[22] and women head of households are usually too poor to afford one. Moreover, the 1978–9 British National

Travel survey showed that only 30 per cent of women had a driving license as compared to 68 per cent of men.[23]

The majority of low-income people travel by bus, but more women than men rely on this mode. In the Nairobi study, 56 per cent of men's trips as against 66 per cent of women's trips were made by bus. In the Belo Horizonte study, 53 per cent of men's trips as against 63 per cent of women's trips were made by bus. Over two-thirds of all women's trips for shopping and medical purposes were made by bus, usually in the off-peak periods.[24]

Following the bus, walking is the next most important way for low-income women to make their trips. In the Nairobi study, 27 per cent of female-headed households, compared to 15 per cent of those headed by men, undertook their trips as pedestrians.[25] In the Belo Horizonte study, 21 per cent of women and 18 per cent of men walked to work.[26] The decision to walk will depend on the availability of transport, the cost of transport and the distance of destination. Little is known about the trade-offs which low-income women make in their use of transport modes. In fact, because of the constraints of time and cost, they may not have much of the 'modal choice' which planners often assume is available to them.

Assignment of routes

Once information about trip generation, modal split and trip distribution has been established, it is used as inputs into the route assignment model, the final phase in the 'standard package'. The exclusion of the majority of women's trips from the first three phases automatically means that their needs will not be considered in the final phase. This is one reason why the routing of public transport is often poorly designed from the point of view of low-income households and low-income women.

Peripheral low-income residential areas are poorly linked to the main transport routes and to places of employment. Therefore, travelling to work by bus – the main mode of travel for low-income women and men – means walking long distances to bus stops and experiencing long delays. For example, in Recife and São Paolo, Brazil, walking to the bus stop accounts for about 25 per cent of total travel time in the work trips of low-income people.[27] In Belo Horizonte, it was found that women had longer trips and had to change modes of transport more often. On average, women spent 60 minutes getting to work as against 45 minutes for men. Women had to take two buses while men needed one. The mean number of stages in women's trips (including walking to the bus stop and waiting for buses) was five as against four for men. Women also spent 34 minutes waiting for buses compared to 24 minutes by men.[28]

Since women make a higher proportion of their trips by bus than do men, the distance from bus routes and the number of bus changes are particularly important to them. Moreover in their non-employment-related trips, they may be trying to combine several errands, such as shopping, accompanying children or older women on clinic visits, etc., and these factors may impair the extent to which they can perform their other jobs effectively and easily.

The scatter of urban activities and the gender differences in the use of these activities makes more gender-aware routing of public transport a complex task. The coordination and control of the task is made even more complex by the dominance of private operators in 'public' transport in many countries. For example, in Lima, Peru, 'private operators fill their vehicles to maximum volume and refuse to plan their routes to feed into lines served by the municipal buses'.[29]

The cost of public transport

In many developing countries, urban-transport planning has not addressed itself to the issue of urban poverty. In many urban areas the so-called 'public' transport sector is run largely by the private sector and the costs of public transport have been passed on to its users – predominantly the low-income people – of whom women comprise the largest proportion.

In areas where fares are based on distance, the fare structure is disadvantageous to low-income households who live on the peripheral urban areas in developing countries. Low-income women are further disadvantaged because, as was suggested earlier, they often work as domestic servants in high-income residential areas or in other informal-service sectors and may have to travel further than men. The Belo Horizonte study shows that women undertook longer trips than men. The structure of fares also tends to make women's multiple stops on multipurpose trips more costly.

In the 1980s the Greater London Council (GLC) made a number of attempts to introduce new and reduced fare structures ('Fares Fair', 'Just a Ticket'). However, the largest reductions were made on the longest journeys on the Underground. Since women's trips tended to be short and were undertaken by bus rather than Underground, they did not benefit as much as men.[30] This policy was introduced around the time the preliminary research for the Women and Transport Survey was underway and highlights the need for quantitative information on women's travel so that the impact of policies can be adequately assessed. Since the study, women have benefited from lower off-peak zonal-return tickets and from the introduction of travelcards for daily, weekly, monthly and longer periods which can be used on both bus and Underground. This gives them more flexibility for multipurpose trips at a fixed fare in both peak and off-peak hours.

Conditions of travel

The conditions of travel on public transport are not designed to suit the needs of women. First, the design of buses make it difficult for women to negotiate public transport accompanied by children, shopping or merchandise for sale.[31] The situation is made worse by overcrowded services. When women are carrying produce or goods made at home to the market, they are forced to wait for less crowded buses. For example, Kenyan women in the village of Mraru found that the infrequency and the resultant overcrowding of buses to the nearest market town, 12 kilometres away, caused severe difficulties in getting their produce to the market.[32]

Second, the fear of violence or harassment in public transport is a very real deterrent to women. For example, a study on low-income women in two shanty areas in Lima, Peru, showed that they commonly experienced sexual harassment in crowded buses and at bus stops, as well as theft, against which 'women travelling with sacks of merchandise, their market baskets and small children, are powerless to defend themselves'.[33] The GLC Women and Transport Survey also found that women were apprehensive about security.[34] Sixty-three per cent of women said they would not use public transport at night on their own. Women expressed concern about being attacked at lonely bus stops while waiting for buses, particularly at night. Increasing automation of the Underground, with the resultant reduction in staff on trains, platforms and stations, also worried the women in the London Survey.

While fears about safety and harassment are complexly related to wider social attitudes and conditions, physical design in transport provision can be much improved. The location of bus stops and the design of subways or bridges can be done in such a way that these are areas of 'defensible space': that is, within earshot and informal surveillance of people. More formal means of surveillance, such as the presence of staff or closed-circuit television, can also serve to reassure women about their safety when travelling.

Community involvement

The issue of community participation in urban-transport planning is still relatively new in many developing countries. Rather, demands for consultation in both developed and developing countries have come from outside the transport-planning discipline, for example from the environment lobby.

Nevertheless, transport has always been an issue around which people have mobilized in urban areas, for example to protest against rising bus fares. In low-income areas, women often take the prime responsibility of organizing neighbourhood groups in order to put pressure on local authorities. In Guayaquil, Ecuador, as the swamps became increasingly occupied by housing, it was the women who urged their neighbours to form self-help committees.[35] Catwalks had been built to traverse the swampland, but these proved treacherous for children and created enormous difficulties for women carrying water, shopping, etc. Infill to make the roads was the first priority. Among the tasks of the committee was to mobilize the community and petition and lobby the local authorities for infrastructure. Regardless of the composition of elected officials, it was the women who were always the overwhelming majority of the rank-and-file members and who took responsibility for the day-to-day work.

In the face of inadequate transport provision by the public sector, women may mobilize to provide their own transport. However, such cases are rare as the investment required can be problematic for low-income women. In the village of Mraru, Kenya, where the local transportation was completely inadequate for the women's domestic and productive needs, the women organized the purchase of a bus.[36] The bus proved so economically successful that the women invested the profits in a shop. Unfortunately, after four and a half years of use, the group found it difficult to replace the bus because of substantially higher prices.

Employment in the transport sector

Employment in the transport sector has traditionally been the domain of male workers. However, the realities of economic transformation, coupled with changing attitudes and opportunities for women, are challenging this traditional gender division of labour. For example, in the UK, although men are more likely to have driver's licences, thus making them eligible to apply for driving jobs, more women are now taking driving tests and being trained for driving jobs in the public transport sector.

The employment of women in the construction of urban-transport infrastructure is another area for further research. Road-building is usually not viewed as 'women's work'. However, in rural areas in many countries, women are being hired to construct and maintain roads. For example, between 1975 and 1982 in Kenya, 30 per cent of the labour force recruited for the construction of rural access roads were women, many of them heads of households. Because of the scarcity of male labour, women were actively recruited and paid the same wages as male road-builders.[37] In Botswana, female labourers were given equal opportunities to train as gangleaders and supervisors on the rural-access road programme.[38]

It is important to distinguish between those road programmes that recognize women's productive role and those that do not. For example, in Lesotho, women were paid for their road maintenance work in food, not cash.[39] This indicates that they were perceived only in their reproductive role – as mothers responsible for feeding their families – and not as producers. With the appropriate gender-aware strategies, the transport sector could provide a variety of employment opportunities for women.

CONCLUSIONS

It has been shown that women's travel needs have been overlooked, basically because of the gender assumptions held by transport planners. The limited work done on women's travel patterns shows that there are great differences between the travel patterns of men and women. It would be a mistake, however, to confuse women's current travel patterns with their actual travel needs. Because women currently travel using transport systems which are not concerned with their particular needs, they have to adapt their travel to the possibilities allowed by the available transport. Thus, the possible false equation of women's low travel mobility with low demand for travel, and their resultant exclusion from planning attention, must be challenged. Rather it should be understood that, in part, women travel less because current transport provision makes it difficult or impossible for them to travel more. What we need is an analysis which is able to explain why women travel less than men.

In redefining women's travel needs, we have seen that there is a complex interrelationship between transport and the performance of the triple role of women. Women's overall travel needs can be defined as the ability to combine their three roles within a finite time period. In this way, transport can not only constrain the ability of women to balance their roles, but it can also affect the performance of each role separately. Inadequate transport can affect women's choice of productive work (limiting their access to better work) and constrain their access to services and other urban activities necessary for their reproductive role. In the end, the ultimate impact is on the welfare of the household as well as women themselves.

It is useful to make a distinction between obligatory and discretionary activities associated with the performance of women's roles.[40] This can serve as an analytical tool to highlight how travel time is traded off against time spent on other activities. There are some areas of women's work where the travel time involved cannot be traded off because those tasks are considered essential for the survival of the family: for example, certain areas of child care, personal care, household maintenance or the marketing of home produce. The patterns of obligatory and discretionary activities are different not only for men and women, but also for women of different income groups and stages in the life cycle.

Transport plans and strategies to meet women's travel needs must of necessity be based on multi-sectoral information. At present, perhaps the greatest constraint to gender-aware transport provision is the lack of quantitative and contextual information about women's travel needs. Such information needs to influence the heart of the transport-planning process: that is, the 'standard package' discussed above.

In evaluating the limitations and the potential of transport policies in meeting gender needs, it is also useful to distinguish between the practical and strategic gender needs of women.[41] To the extent that transport policies aim at alleviating women's triple role, both in enhancing the performance of each role and in minimizing the time spent in travel, they are meeting the practical gender needs of women – those that arise from the concrete conditions of their position in the existing gender division of labour.[42] For example, trips for shopping, work, collecting children, etc., arise from women's practical gender needs. In fact, these are needs for all household members. Their identification as 'women's needs' reinforces women's responsibilty for the reproductive work in the household.

Women's strategic needs are those that arise out of their subordination and are formulated in terms of a more satisfactory organization of society in terms of the relationship between men and women.[43] For example, the ability of women in some Muslim societies to travel freely in public is identified as a strategic gender need for these women.

Meeting practical gender needs will not automatically further strategic gender needs. This is particularly true for transport. Going back to the example of Muslim women, even if accessible and safe transport were made available to enable women to combine their triple role in the company of a male relative, it is unlikely that women would freely and independently use these services unless there were prior changes in the attitudes of both the women themselves and the men in the community. Nevertheless, the availability of such transport might play a role in influencing attitudes.

Studies of women's travel time show that women's entry into the workforce has made little change in the gender division of labour in the household. The findings show household chores were still done by women, and not shared with their male partners, but the travel needed to do the chores was deferred to the weekend.[44] Therefore, the aim of integrating gender into urban-transport planning may initially be focused on meeting the practical gender needs associated with women balancing their three roles. In doing this, urban transport plays a critical role in the 'full income' of the household,[45] has important consequences for the survival and the welfare of the household and may influence strategic gender needs in the longer term.

NOTES

1 Dimitriou and Safier (1982).
2 Moser (1989).
3 Moser (1989).
4 Moser (1989).
5 Moser (1989).
6 Fox (1983).
7 Pickup (1984); Fox (1983).
8 For example, Schmink (1982); Anderson and Panzio (1986).
9 See Fox (1983); Pickup (1984); Hanson and Hanson (1981).
10 Thomson (1983).
11 Schmink (1982).
12 Rosenbloom (1978); Fox (1983); Hanson and Hanson (1981).
13 Rosenbloom (1978, p. 348).
14 Schmink (1982).
15 Fox (1983, p. 159).

16 See Fox (1983) and Schmink (1982).
17 Schmink (1982).
18 Schmink (1982).
19 Fox (1983); Hanson and Hanson (1981).
20 Nairobi City Council (1984).
21 Schmink (1982).
22 Pickup (1984); Fox (1983).
23 Pickup (1984).
24 Schmink (1982).
25 Nairobi City Council (1984).
26 Schmink (1982).
27 Schmink (1982).
28 Schmink (1982).
29 Anderson and Panzio (1986).
30 GLC Women's Committee (1984c).
31 See Pickup (1984).
32 Kneerim (1980).
33 Anderson and Panzio (1986).
34 GLC Women's Committee (1985b).
35 Moser (1987a).
36 Kneerim (1980).
37 World Bank (1982).
38 Hagen et al. (1988).
39 Edmonds et al. (1986).
40 Fox (1983).
41 Moser (1989).
42 Molyneux (1985); Moser (1989).
43 Molyneux (1985); Moser (1989).
44 Hanson and Hanson (1981).
45 Schmink (1982).

Chapter 7

Health

Lise Østergaard

GENDER – HEALTH – DEVELOPMENT

Development and health are intrinsically interrelated: without a certain level of economic and social development, we cannot provide the population with basic health care. And without a basic state of health, the population does not have the physical and mental energy necessary to develop the society. It is less widely recognized that in this chicken-and-egg dilemma a third issue is always present: gender. Can a nation give birth to healthy babies without healthy mothers? Will the sucklings survive without sufficient mothers' milk? Will the family gain enough energy from its meagre resources if the food-providing female is not able to produce sufficient provisions and practise good nutrition? Will the growth in population stop exploding if females have no knowledge about and no access to family-planning measures? Gender, health and development make up a dynamic triad.

The concept of health covers a complex human condition. It cannot be approached from a medical perspective alone, nor can it be fully dealt with in society by the health sector alone. For the formal health sector tends to operate with a 'medicalized' concept of health, dealing with circumscribed diseases rather than with the human being in its environmental context. Thus it tends to deal in corrective measures. However, preventive health care is no less important. To succeed, preventive health measures must identify all kinds of pathogenic factors – biological, cultural, economic, social and political – and respond to them through an interdisciplinary and intersectoral approach that involves the total community. In this totality, gender plays a decisive role.

The concept of disease

The concept of disease is more easily defined than the concept of health, and yet the suffering of people does not always fit well into the medical nosology. To sharpen the understanding, we might distinguish between *diseases*, being the doctor's objective diagnosis of abnormal conditions, and *illness*, being the patient's subjective perception of suffering and being unwell. An imaginary disease may cause serious suffering and a serious disease may be unnoticed by the patient. More importantly, a serious condition like a low haemoglobin percentage, which in an affluent country is regarded as clearly abnormal, may be widespread in a population of malnourished, anaemic Third World women, who do not even regard their state of chronic fatigue as being abnormal. More often than not, the sufferings of women are not fully recognized nor correctly interpreted by medical professionals, who are mostly men.

As a consequence of the accelerated development of medical high technology, there is a strong tendency in industrialized as well as developing countries to define health needs within a narrow concept of distinct diseases to be prevented by immunization and cured by drugs or other treatment, rather than defining health needs in terms of those human activities and social structures which are actually carriers of good or bad health. Economic, social and cultural conditions, lifestyle and life stress are the major determinants of health. Yet most countries continuously pour out increasing resources for curative high technology, which usually benefits only an urban élite population. Long-term investments in public health – such as the provision of adequate food and safe water, the elimination of physically undermining workloads and the like – are regarded as being outside the health sector.

According to the 1978 WHO/UNICEF Alma Ata Declaration, health is more than the absence of disease or infirmity, it is a state of physical, mental and social well-being.[1] There is a baseline below which no individuals in any country should find themselves, a level of health that will permit them to lead a socially and economically productive life. Given this, people will realize that ill health is not inevitable: that they themselves have the power to shape their lives and the lives of their families, free from the avoidable burden of disease. Unfortunately, this ideal condition is not equally obtainable for women and men, for the health risks of women are greater, not least during the reproductive years.

The concept of development

Health and development are closely interlinked in the Alma Ata philosophy of *Health for All by the Year 2000*. Health is an integral part of development because people are both the means and the ends of development. The human energy generated by improved health should be channelled into sustainable economic and social development; and economic and social development should in turn be harnessed to improve the health of people.

Measuring development in terms of access to basic services such as health care, food security, safe water and primary education is more informative than using purely economic yardsticks, such as per capita income. Social indicators like infant mortality and life expectancy reflect even more accurately the living conditions of most of the population, because of their broad distribution across households. Kenya, for example, with a GNP of only $330 per capita, has lower infant mortality and a higher primary school enrolment rate than, say, the Ivory Coast with a GNP of $740 per capita.[2]

The concept of gender

By considering gender with health and development, we recognize that health opportunities and health hazards are not the same for men and women. What is commonly accepted as the typical attributes of men and women differ among cultures, societies, classes and over time. In certain cultures, for instance, it has been part of the male image to be strong and capable of bearing pain without complaint, while in others it has been women who are expected to suffer pain in silence. We can imagine that such expectations will influence help-seeking behaviour in times of illness.

Health planners and practitioners, who are themselves often male and Western, may implicitly view the needs for health care and the effects of health projects from a male and Western perspective. They may thus fail to see that the health needs of Third World females and males, whether adults or children, can vary considerably because of gender differences. We still have too little insight into the full effects of gender issues on the status of health. Much research needs to be done. But a first step that can be taken now is the provision of health-related statistics analysed for gender differences. Armed with such data, development planners and practitioners would have far greater prospects of improving the health care of women.

RESOURCES AND HEALTH

Poverty

Poverty is the world's most serious carrier of ill health. In parts of the world, debt crises have led to structural changes in the national economy, followed by savings on civil budgets and unjust policies of income distribution. Such policies tend to make the poor section of the population even poorer; and among these women are most adversely affected. In the Third World there seems to be a widening gap between the income-generating ability of women and that of men. Paradoxically, development has not helped to improve the status and health of poor women, but rather has had a negative effect. We are dealing with a complex of interacting dynamics, mutually reinforcing each other with an aggravating effect.

A band of vicious circles spins through peoples' lives:

- Poverty
- Malnutrition
- Chronic diseases
- Increased reproductive strain
- Fatigue and apathy
- Lack of education
- Poor income-generating ability
- Leading to increased poverty.

The carriers of ill health may also be broken up into smaller circuits, as for instance:

- High reproductive activity
- Poor health
- High infant mortality
- Leading to increased pressure on reproduction.

Or:

- Low socio-economic status
- Low self-esteem
- Reluctance to seek health care
- Chronic diseases
- Still more apathy
- Leading to still lower self-esteem.

Reducing the effect of just one of the factors may ease the situation, but influencing several of the elements in these dynamic interplays at the same time may break down the vicious circle. The health situation is definitely improved, for instance, by immunization against diseases, in turn improving energy and the capacity for work. But immunization programmes alone are 'vertical' by nature, because they only isolate a single factor of the complex nature of health.[3]

The mutually reinforcing dynamism of comprehensive resource-development programmes is necessary in order to turn these increasingly negative, vicious circles into the positive spiral of socio-economic development, improving health and in turn liberating energies for further development. Family planning, for example, is more easily accepted when overall mortality is relatively low and levels of education relatively high. In turn, the spacing of births improves maternal and child health and further limits the degree of mortality. Similarly, clean water and sanitation produce more benefits when provided together with health education, which together with improved dietary habits reduces intestinal infections in children and thereby boosts nutritional status. Healthier children are more likely to attend school and to learn than the sick and malnourished, and in turn education enables people to understand health problems and to act towards their prevention. Basic health care is an essential key to open the door to sustainable development.

Food security

Food security has deteriorated in Africa south of the Sahara and severe food shortages are now widespread. Recurrent famines have illustrated the high degree of food insecurity in the region. Between 1965 and 1986 food consumption in sub-Saharan Africa averaged about 85 per cent of recommended requirements.[4] With a highly uneven distribution, the consumption of hundreds of millions of people is far below that level. Within poor households, women and children are more susceptible to malnutrition and are therefore a specific high-risk group in relation to health.

Nutritional deficiencies are known to have strong interactions with diseases, as they lower the body's immune response. The infections tend to be more severe and the duration of illness longer, contributing to further deterioration of the person's nutritional status.

In many Third World countries, malnutrition is seasonal and increases before the harvest when food supplies have dwindled. The seasonal nature of agriculture brings special problems to women. In The Gambia it has been noted that pregnant women loose weight during the peak agricultural season; in Thailand there is a marked increase in miscarriages as well as early termination of breastfeeding during rice planting and harvesting. The women do not have the time to prepare the necessary weaning foods and may only cook and eat once a day. Children are born low-weight for age and therefore are at high risk for morbidity and mortality.[5]

Even when sufficient supplies of food are available at the national level, there is no guarantee that they will reach those who need it at the household or individual level. Sometimes this results from inadequacies in infrastructure and transport, but more often it results from the fact that those who need it most do not have the means to buy it. Green revolutions have achieved self-sufficiency of food for countries but not for individuals and poor families. It is at the level of the individual and the household that the challenge to the achievement of sustainable food security must be met.

The most direct way to counteract malnutrition would simply be the encouragement of an expanded production of what the poor traditionally eat, such as millet, other coarse grains and root crops. These are cheap sources of calories and can be produced by small farmers, including female heads of households, directly serving the nutritional needs of the family. Today, however, concern is given more to market forces than to nutritional needs. High-technology monocultures are supported and low-technology subsistence farming or farm-household production is neglected at the cost of nutrition for numerous families.

A reorientation in international as well as in national planning towards emphasis on crops favoured by the poor and grown under conditions faced by the poor could improve the situation at very low cost. It is a drawback that nutrition policy and food policy are most often treated separately at a national level as well as in international organizations. Incorporating nutritional concerns explicitly into rural programmes would greatly increase the benefits of agricultural planning and could neutralize its possible negative effects.[6]

Water and sanitation

Unsafe water and poor sanitation are two other direct derivations of poverty. They are serious carriers of ill health and another area where gender issues are crucial. The provision of water for households is the duty of women in most of the world where water has to be carried. Women sometimes spend as much as six hours a day walking long distances and carrying heavy loads of water. This arduous task exploits their energy and time; and since they are often nutritionally vulnerable, it further endangers their health. In fact, the carrying capacity of the women and girls in a family and the distance to the well are the decisive factors determining the actual supply of water. If the capacity is limited and if the quality of water is poor, the level of hygiene will be low and many waterborne diseases, infectious as well as parasitic, will occur. Their daily and direct contact with water makes women particularly susceptible to water-related diseases.[7]

Because of its association with women, the provision of water for household purposes is an undervalued issue and appropriate technology to alleviate this burdensome task is lacking. Although women are the providers and main users of water, they are seldom consulted when it comes to the initiation of water projects. Male engineers and administrators usually believe that women are incapable of managerial roles in relation to water and sanitation, in spite of the fact that they are traditionally the ones responsible for these resources in and around the home. Women have the incentive to make water programmes work, since they are the most affected by poor access to water. Wherever communities are involved in the design, construction, installation and maintenance of water supplies, water projects are more efficient, cost-effective and hence sustainable.[8]

WHO estimates that three out of four people in the developing countries do not have access to safe drinking water and only one out of four has access to sanitary facilities. WHO further estimates that 80 per cent of all diseases are related to unsatisfactory supplies of water and sanitation. Diarrhoea directly kills millions of children every year and contributes to malnutrition. Diarrhoea and poor nutrition form a vicious circle, since malnourished children are more seriously affected by diarrhoea, which in turn further deteriorates the nutritional condition.

Parasitic worms infect nearly one half of the entire population of the developing countries, often with very serious consequences. For example, 200 million people in 70 countries suffer the debilitating effects of schistosomiasis. Trachoma affects some 500 million people at any given time, often causing blindness. Each year malaria kills about three-quarters of a million children aged under one year in sub-Saharan Africa alone.[9]

It is imperative that development planners and practitioners repair their long neglect of household and family needs for water supplies. Far too often water schemes are macro-installations serving fields for export-cropping, while the micro-installations serving the nearby villages are not implemented. Or when water schemes do supply villages, the necessary arrangements for maintenance are not made.

Workload and industrial injuries

Overwork adds to the health risks of poor Third World women. Time–budget studies have recorded the extremely long hours that women work, showing that 15 hours a day, seven days a week, all year round is not unusual for poor women in places as far apart as Burkina Faso and Northern India.[10]

Women produce as much as 60–80 per cent of subsistence food in Africa and more than half of the food in the least-developed countries all over the world. Yet rather than easing their task or boosting their productivity, the so-called development of agricultural methods has often increased the burden on women food producers through loss of land for export-cropping, leading to an increase in their traditional labour-intensive responsibilities. For example, the combined effects of this loss of land for export-cropping and population pressure have marginalized formerly productive small-scale farmers and made them dependent on an increasingly precarious access to cash income in order to buy food. This in turn has led to a massive rural exodus – predominantly a male migration to cities – which has left the women alone in rural areas, responsible for the survival of their families. In urban slums, without support from the family network, uneducated and unskilled women are employed as casual labourers, vendors or domestic servants on very low wages. Many are petty traders or turn to prostitution. The lack of day-care centres means that mothers are forced to leave their children in the care of older siblings or alone by themselves.The UN reports that about one-third of all households in the Third World are now headed by females.

The normal work of poor Third World women exposes them to an enormous number of direct health risks. Studies in India show, for instance, that the amount of cancer-causing particles breathed in by women cooking over open, smoking stoves in ill-ventilated houses is equivalent to smoking twenty packets of cigarettes a day.[11]

The increasing use of pesticides in agriculture is creating new health hazards for agricultural labour. While the chemicals are intended to be handled with extreme caution, precautionary measures are usually not adopted in many developing countries. The spraying of pesticides is carried out by untrained persons, most often women, with no safety devices. Accidents from the use of sub-standard agricultural machinery

– mainly threshers – are also being increasingly reported. How is a poor, rural female-headed household to survive with a maimed mother?

To supplement their incomes, poor households engage in 'cottage-industry production'. Often this work is carried out in the home for extremely low piece rates. Some of this piece-work is a severe strain on the eyes, such as very fine embroidery. The rolling of indigenous cigarettes (bidis) are associated with tuberculosis and other respiratory tract infections because of poor domestic working conditions.

The electronics industry has perceived the economic advantage to be gained from combining high-technology with very labour-intensive component assembly. Female labour is preferred here because of women's 'nimble fingers' and great patience. But, as in the modern sector in general, women are perceived as marginal labour. Since the women have not yet been able to unionize, they are underpaid and exposed to health-destroying working conditions, performing repetitive low-skilled labour tasks in dusty, overheated quarters to extremely high production standards, working virtually round the clock during periods of peak production. Female labour is constantly supervised to the extent that extremely brief toilet visits are regulated and controlled; meal breaks are so short that it is virtually impossible to finish eating. Women are forced to make a choice between continuing to work and losing their health, or quitting and losing their livelihood.

Education

Education is intrinsic to development. In the widest sense, empowering people with basic cognitive skills is the surest way to render them healthy and self-reliant human beings. Research has also established that a mother's education enhances the probability of child survival. So, to raise health standards, we must also raise education levels.

In the less-developed countries there is a disparity between male and female literacy. Generally two out of every three illiterates are women. In some areas nine out of ten women are still illiterate. Furthermore, female illiteracy is three to four times higher in rural than in urban areas. The chance of little girls having seats in primary classes is very small; and when they get seats, they are not likely to keep them very long. Where admittance of girls to school is strictly limited, the drop-out rate of those who actually get started is proportionally high. For a vicious circle operates in regard to women; because of overwork, women seek the help of their daughters, which deprives the girls of schooling and access to literacy, whereby they in turn are handicapped in relation to vocational training.

Under certain circumstances girls are not sent to school because their parents do not expect to benefit from it. Daughters are not expected to support themselves or their aged parents later in life and any benefit that does arise from education will be reaped by their marital households. Female illiteracy has been described by UNESCO as an endemic problem in three-quarters of the world.[12]

The gender gap in education comes at a high cost. Evidence shows that the mother's education is the single most important determinant of a family's health and nutrition. Further, even a few years of primary school have been shown to lower women's wish to bear many children either directly by increasing awareness of contraception or indirectly because of increased access to their own economic resources.

Education is a catalyst operating behind all the carriers of ill health previously mentioned. Lack of education aggravates their effects, whereas sufficient education alleviates their most devastating consequences. Education is a means of overcoming poverty, increasing income, improving nutrition and health, reducing family size and, not least important, raising people's self-confidence and enriching the quality of their lives. Educating girls is the best investment a country can make in future economic growth and welfare, because of women's almost exclusive influence in the home on health, nutrition and fertility, and because of the formative influence of mothers on the next generation.

DIFFERENTIAL PRACTICES AND HEALTH

Son preference and daughter neglect

The preference of sons is based on purely economic grounds. The birth of a son signifies the continuation of the family, a source of income and marriage and financial security for the parents in their old age. But the birth of a girl is most often regarded as a financial burden, unless several sons have already been born. 'To get a girl is like watering the neighbour's tree, you have trouble and expenses in nurturing the plant, but the profit goes to somebody else.'[13]

Particularly where resources are scarce, the son preference in families necessarily leads to the neglect of daughters. In the most extreme cases it leads to the abandonment of the baby girl or even to female infanticide. This low status produces a minimum investment in females from childhood and throughout life. The result is the low-health status of Third World women characterized by low-life expectancy, high rates of mortality and chronic low-grade morbidity.[14]

But, in areas where productivity is high and at times when there is high demand for female labour, discrimination against women is low. This mechanism acts even with children, to the effect that where women's expected employment is high, little girls receive a larger share of the family resources and are more likely to survive.[15]

Differential feeding

The effect of son preference has a mental as well as a physical impact on discriminated daughters. It affects their self-esteem in an adverse way and socializes them into putting themselves last when the distribution of food and other household resources takes place.[16] The discriminatory feeding of females starts right from early childhood. In many countries girls are weaned substantially earlier than boys and generally fed less well than boys, giving the girls significantly smaller chances of surviving and harming their developmental potential to a degree that may be dramatic when they themselves, perhaps as teenagers, are to give birth. In the general food-distribution system in families all over the world, males are given priority over females and elder family members over younger. Food is distributed in proportion to the prestige of the family member rather than in proportion to biological needs. If food distribution were based on an understanding of and concern for the biological needs of the family members, the priority would be different. The needs of small children, girls as well as boys, would be given precedence and young girls preparing for reproduction as well as pregnant and lactating mothers would be served well enough to let them carry out their reproductive functions with less waste of life than is now the case.

A study carried out in a group of villages in Bangladesh found that up to the age of five years, the calorie intake of girls was on average 16 per cent less than that of boys; the discrepancy was 11 per cent for the age group 5–14 years. As a result, 14 per cent of the female children in the survey were severely malnourished compared to 5 per cent of the male children and 26 per cent of the girls were severely stunted compared to 18 per cent of the boys. Women who are anxious for sons tend to stop breast-feeding the baby girl as soon as they become pregnant again. If, on the other hand, the first-born child is a son, they try to delay another pregnancy as long as possible and breast-feed for an extended period to give the baby a sound start in life.[17]

The sad results of these habits are documented by a number of objective observations. Studies from different parts of the world report gender differentials in nutritional status as indicated by anthropometric measurements. When social variables are compared, the most significant determinant of nutritional status is found to be sex.[18] Significantly more girls than boys are malnourished and have been given less of the food resources available from a quantitative as well as from a qualitative measure.

In addition to these disadvantages, young women are in many places exposed to dietary taboos during pregnancy, with strong negative effects on their health and the health of the foetus. There are reductions in diet (particularly the exclusion of fish, eggs, meat and other high-protein nutrients), general reductions of food intake, or the exclusion of specific food items for fear of having an over-large foetus with an over-large head, causing difficult delivery. These practices and fears are decisive in countries where the malnourishment of mothers and the low birthweight of infants is common. Women's fear of having a large baby may most of all reflect the sad reality of poor care during childbirth.

Differential health care

The son-preference/daughter-neglect syndrome also shows up in poorer health care for girls during illness. In spite of the fact that there are more malnourished girls than boys, statistics show that more ill boys are brought to the health clinics while girls are kept at home. When health visits are paid to the villages, a substantially higher number of ill and malnourished girls are seen. The same picture is seen in relation to immunization if a small fee is charged. If immunizations are not given free, the proportion of girls brought to clinics is about one-quarter that of boys.[19] Worst off are the late-born female children. A comparison of sex and birth order with medical expenditure shows a virtual absence of medical treatment for late-born females.

Generally speaking, girls are born with a biological advantage, which makes them more resistant to infection and malnutrition than boys. So, where treatment is even-handed, girls should be at less risk of dying in their first five years of life than boys. The odds are 1.15:1 in favour of girls. But in a number of countries and on every continent the biological advantage of girls has been cancelled out by their social disadvantage.[20]

Such attitudes and habits, of course, have far-reaching health implications and damage the health of females considerably. The ultimate consequence is an excess female mortality in childhood, which indicates that serious external influences act against the normal biological advantages of girls compared to boys. These adverse influences are partly the exposure to risk of health impairment, partly the ill effects of poor or lacking treatment. Excess female infant and child mortality should therefore be seen as a warning signal, indicating serious neglect of girls in the society concerned.

Unfortunately, it is very rare for gender differentiated health data to exist together with field studies on the prevalence of discrimination by sex. However, female child-mortality rates higher than those of males are reported in a number of studies.

It is important that planners of health care are aware of such hidden attitudes of daughter neglect and take measures for the prevention and abolishment of these injurious practices. Prolonged action is necessary, for we are up against entrenched attitudes which are not easy or quick to change.

Nevertheless, various preventive measures are possible. A first step are relevant observations to identify whether the problem exists in a given country. This can be followed by information to parents and steps to have the girls included in the existing health services, whether preventive or curative. Information about alternative food provision and the special nutritional needs of pregnant women and lactating mothers is needed, along with training of parents in the treatment of malnourished children: for instance, those suffering from kwarshiorkor. In some cases short-term special programmes to protect the neglected girls may be helpful, although this is not a permanent solution. The creation of an awareness among health and social workers of the risks run by female children in the societies where discrimination takes place is equally important.

REPRODUCTION AND HEALTH

High fertility and infertility

In developing countries, women's social situation is strongly determined by family structure and motherhood is given a high social value. Children are wealth: having children gives women status and respect, so women are proud of their child-bearing capacity. As women have authority within the sphere of the home, children increase the scope of their activity and power. Men take pride in fatherhood and confirm their reputation for virility by having many children. So a wife who bears many children increases her own influence and prestige as well as the prestige of her husband.[21]

In addition to being valued for their own sake, children are valued for the help and care they can provide for the family. They begin to work at an early age inside as well as outside the household and when older they are the only form of old age security their parents have.

However, the culturally determined preference for many children also causes severe disadvantages for women by exposing them to social pressures to bear many children, especially those of the 'right' sex. In some countries, like India, where for cultural reasons parents may only depend on their sons, there is a strong preference for male children. Many women will continue trying to give birth to a son no matter how many daughters they have in the meantime. The need for one or two sons to survive into adulthood may force the woman to a high number of childbirths.

Yet high fertility goes hand-in-hand with high infant mortality, increasing the pressure on women to have more children. If the odds were greater that children would survive, few women would choose or need to become pregnant some fifteen times and give birth to eight or nine children when their lives are already strained by so many other burdens. The combination of higher income, better health, more education and a growing acceptance of family planning have begun to reduce birth rates in most middle-income countries.

The other side of this coin is infertility, which is a curse in the Third World; it deprives women of any social value. Women who cannot bear children are considered worthless; they have to face rejection from their husbands and family and may be exposed to divorce or desertion. In parts of the world this is also true for women who bear only daughters. The low status of women permits people to assume that it is always the woman who is infertile as well as responsible for the sex of the child.

Maternal mortality

Some 500,000 maternal deaths are estimated to occur each year; primarily in poor countries, among poor women and often in distant villages. For these reasons exact statistics are not available; the experience of experts is that maternal deaths are virtually always underreported. Data from Egypt, India, Indonesia, Malaysia and Turkey, gathered by a 1986 WHO report, showed that large proportions of maternal deaths took place either at home or on the way to the hospital.[22] These proportions ranged from 24 per cent of deaths in Turkey to 82 per cent in rural India. In Bangladesh, hospital staff were aware of only 4 per cent of the maternal deaths discovered by researchers.

The health risks for child-bearing women in the Third World are enormous. Local statistics show, for example, that each time women in rural Bangladesh become pregnant they face a risk of dying which is at least 55 times higher than that faced by women in Portugal and 400 times higher than the risk for women in Scandinavia.[23] Because of the high birth rates in the developing countries, the lifetime risk of a woman dying in pregnancy or pregnancy-related illness is 1 in 25 or 1 in 40, which contrasts sharply with the 1 in 3,000 for women in the developed world. In addition to the hundreds of thousands of women who die in pregnancy and childbirth, millions more are permanently disabled and many of them are ostracized by their families and communities. Yet it is estimated that 63 per cent to 80 per cent of direct maternal deaths and 88 per cent to 98 per cent of all pregnancy-related deaths in the Third World could probably have been avoided with proper handling.

The causes of this high degree of morbidity and mortality are many and complex. One is too many childbirths – too early in life, or too late, or too close together. No less decisive are poor obstetric care, the fatal consequences of illegal abortions, the detrimental results of female circumcision and the generally poor physical condition of the undernourished and overworked mothers.

The safest period of a women's life for childbearing is from age 20 to 30, yet between 10 per cent and 20 per cent of babies born in developing countries are to women in their teens who may be little more than children themselves. Because their bodies are not yet fully prepared for the demands of childbirth, teenagers stand an excess risk of death compared to women aged 20 to 24 years.[24] In a study in Nigeria, for example, women aged 15 had a maternal mortality rate 7 times higher than women aged 20 to 24.

Teenage marriage is another reflection of the low status of women. Widespread in the developing world, it has the highest incidence in Bangladesh, where 72 per cent of women aged 15–19 are married. For south Asia as a whole, the percentage is 54 per cent, compared to 24 per cent for southeast Asia, 44 per cent for Africa and 16 per cent for Latin America. Teenage marriage is intimately tied up with the image of ideal womanhood and the need for men in male-dominated societies to control female sexuality. Furthermore, where daughters are considered a financial burden, who cannot earn for their own family nor look after the parents in old age, there is every incentive to be free of that burden as quickly as possible.[25]

Older women also face greater hazards from childbearing. Comparisons of six studies show that for women aged 35 to 39, the relative risk of dying from a given pregnancy varies from 1.85 to 5.61 compared to women aged 20 to 24.[26] Other studies confirm the increased risk of death associated with having many children. In Jamaica, for example, it was found that women having their fifth through ninth birth were 43 per cent more likely to die than women having their second child. In Portugal, women having their fifth birth were three times as likely to die as women having their second, while women having their sixth or later birth were at even greater risk.[27]

Maternal morbidity

Where socio-economic conditions are poor, women are most vulnerable to the health risks associated with bearing children in quick succession. Pregnancy and breastfeeding make big nutritional demands that women from poor homes are seldom able to meet, either by eating more or by getting more rest.[28] Undernourished women are more likely to have spontaneous abortions, still births, or babies with low birth weights (below 2.5 kg) whose chances of survival are thereby reduced. For example, one study found that among Indian women in the low socio-economic groups subsisting on diets providing less than 1,850 kcal and 44 grams of protein daily, pregnancy wastage due to abortions and still births was around 30 per cent.[29]

Most deliveries in developing countries continue to take place at home under unhygienic conditions, attended by relatives and/or traditional birth attendants (TBAs). Several governments have recognized the importance of raising the skills of TBAs and have launched training programmes with this in view. The role model for training TBAs is the Western midwife, and covers the entire span from antenatal to postnatal care including advice on breastfeeding, weaning and family planning.

In many cultures, however, birthing is considered an 'unclean' or polluting process; and in spite of her access to families at all social levels, the TBA has a low social status. In India, for example, she usually belongs to one of the lower castes; because of her social status she is not able to perform many of the tasks expected of her by the training programme, including that of being an agent for changing habits. It is therefore necessary that the trainers understand the setting in which the TBA is operating and then take steps to ensure that she has the necessary status as a trained person within her setting, so that the community recognizes the value of both using her services and of broadening her role. This has been done successfully in some countries, including Botswana.[30]

Most important of all, however, is the provision of improved obstetric care. Networks must be established in the primary-health-care systems which can secure prenatal care and referral to health clinics and hospitals for delivery in case of perinatal complications.

Other factors

Other factors also contribute to maternal morbidity and mortality. Illegally induced abortion is a major killer of women and is responsible for up to 50 per cent of maternal deaths. However, reluctance or inability to get formal medical care results in selective underreporting of these abortion deaths.

Female circumcision, practised in parts of Africa, is another hazard to women's health. It can result in serious medical consequences such as infections of the vagina, urinary tract and pelvis, followed by infertility or problems in childbirth.[31] Yet another health hazard is the increasing incidence of sexually transmitted diseases, which are not easy to detect in women and therefore not well treated. In addition to causing infertility, these diseases may cause continuous low-grade morbidity and fatigue in the woman and can infect the babies. The consequences of AIDS today cannot even be measured.

Family planning

In some parts of the world, women are deprived access to family-planning measures for religious or other ideological reasons and are at the mercy of backstreet abortionists. In other parts of the world, family planning is a euphemism for population control and people are persuaded or forced to submit to irreversible sterilization methods. Both these situations reflect a violation of essential human rights.

With the issue of birth control a very delicate balance must be found and preserved if the human rights of women are to be protected. It should be women's right to have free access to family planning so that they are in command of their own fertility. Moreover, they must be properly informed about the eventual side-effects of the contraceptives they may use. This is of paramount importance if the question is about female contraceptive surgery, where the impact on fertility is irreversible.

On the other hand, it should be an equally unquestioned right for women not to be viewed and dealt with by authorities – including donor agencies – as 'objects whose fertility is to be controlled'. Women should be seen as individual human beings who love and enjoy their children if their living conditions are reasonably tolerable and who want to control their own fertility. I think we should be careful not to become too technical about this issue. Family planning must be a matter of female choice and women should actively participate in decisions about their own reproductive functions.

Because family-planning programmes have been highly controversial in parts of the world, the ethical questions they raise must be addressed by development planners and administrators. Family-planning programmes should never be executed in isolation, but always together with and preferably following after other health services, because we know that no female contraceptive is totally safe.[32] Even a diaphragm is dangerous and may cause infections under poor sanitary conditions. Oral contraceptives should never be distributed without a general medical check and without proper instruction and control that the instruction has been fully understood. Intrauterine devices should never be applied without proper control afterwards for side-effects or complications.

In a number of countries contraceptives are distributed without such precautionary measures, often by poorly trained paramedical personnel. This is done with the justification that the high birth rates do not allow us to wait for an improvement in the health infrastructure. It is also claimed that the risk to health from an additional pregnancy is higher than that from the use of the contraceptives. But it is not acceptable that women should be forced to choose between two evils. High priority must be given to integrate family planning into a general health-care network and women must be properly informed about the various risks before they decide to adopt one of these methods.

Furthermore, research on contraception and family-planning campaigns are almost exclusively directed towards women and not towards men. It is as if men were non-existent and therefore non-responsible for human fertility. The responsibility of fatherhood should be upgraded to equal the responsibility of motherhood. Since no female contraceptive is totally safe and since the condom is the only safe contraceptive known until now, the practical strategy is to urge men to use condoms and to help make them available and cheap. Men should be persuaded to use condoms not only in extramatrimonial relations to protect themselves, but also in matrimonial relations to protect their wives against unwanted pregnancies which may eventually lead to dangerous abortions.

Poverty and fertility

Economists and demographers have argued that continuing or rising birth rates, combined with the higher population growth which follows falling crude death rates, will retard economic development and perpetuate poverty. Based on this, governments, sometimes urged by aid-agencies, have introduced isolated fertility-control programmes, particularly in rural areas. In some cases, fertility-control measures are imposed upon women, either directly or indirectly, in order for them to receive general health services or other privileges. Incentives and disincentives in the form of economic rewards or punishments are sometimes used to promote a particular method of contraception or a certain family size. Such schemes always work in the direction of forcing the most disadvantaged to comply with authorities without any real choice. Moreover, they do the health service serious harm by accepting the concept of conditionality.

Setting incentives for health and family-planning personnel, together with targets for the recruitment of family-planning acceptors, will inevitably influence the way personnel prioritize their tasks and can lead to the neglect of people's basic health needs. The Indian family-planning programme, for instance, is popularly known as the sterilization programme because the highest incentives – and thereby staff emphasis – are given for sterilizations.

In the vast body of literature on fertility there is strong evidence that isolated measures to control fertility have never led to a lasting reduction of birth rates. Human fertility is determined by far more complex clusters of social determinants, most of which are to be found outside the health sector. These include the assurance that our children will survive, the availability of some form of old age security and an improvement in living standards so that children's labour is no longer a necessity, and parents can afford to support them through their childhood and invest in their education with the perspective of giving them a better life.

It is not the high birth rates which create poverty but poverty itself, with the associated conditions of mortality and morbidity, which creates high fertility. In parts of India, such as Kerala, after social and economic changes had taken place the birth rate was much lower than for the rest of the country.[33] Among those sections of the Indian population who have some form of social security as well as aspirations for themselves and their children, for instance permanent employees of government and industry, family size is also shrinking.

In countries where government initiatives are taken to integrate family-planning programmes with maternal and child health care, for example Kenya, surveys indicate a notable increase in the use of both modern and traditional methods of family planning. In 1977–8, 7 per cent of married women in Kenya were using a contraceptive method compared to 17 per cent in 1984. That study found strong indicators of underlying demand and unmet need for both permanent and temporary methods of fertility regulation.[34] With social and economic changes – such as increasing pressure on land, the tremendous burden of school fees, etc. – women realize that they have to adopt new strategies to survive and consequently want to limit their childbirths. However, those desires often meet resistance from their husbands; and the women frequently practise family planning without their partners' knowledge.

HEALTH SECTOR AND HEALTH NEEDS

Development of the health sector

Many developing countries were formerly part of colonial empires and attained political independence only in the second half of this century. The health-service systems they inherited had been set up mainly for the use of the colonial establishment (the administration, public services and the army) and consisted of a few hospitals in the Western medical tradition located in the urban areas where this section of the population was concentrated. The vast majority of the population, however, lived in the rural areas and was served mainly by traditional healers and practitioners of indigenous systems of medicine.

When the newly independent national governments were faced with the task of setting priorities and developing national plans, improved health status for the people was generally considered a desirable goal. This was to be brought about by expanding the health infrastructure, including training personnel in modern Western medicine. But because of the aspirations of the political élite and the medical profession, the lion's share of the resources still went into the urban areas.

The initial expansion of the health sector into the rural areas in the early 1960s and 1970s was often financed by funds provided by aid agencies which were primarily interested in the promotion of family-planning programmes. They were therefore more concerned with family limitation than with providing basic curative care or dealing with the underlying causes of illness.[35] It was the limited success of these programmes which led to the incorporation of maternal and child health care, thereby addressing women's reproductive role more fully.

Primary health care

Realizing the complexity of people's health needs and the limitations of a purely medical health sector, the concept of Primary Health Care was launched at the WHO-UNICEF International Conference in Alma Ata in 1978.[36]

The Primary Health Care programme contains eight essential elements:

1 Education concerning the prevailing health problems and methods of preventing and controlling them
2 Promotion of food supply and proper nutrition
3 Adequate supply of safe water and basic sanitation
4 Maternal and child health, including family planning
5 Immunization against the major infectious diseases
6 Prevention and control of locally endemic diseases
7 Appropriate treatment of common diseases and injuries
8 Provision of essential drugs.

The Alma Ata declaration stresses the right and duty of people to participate individually and collectively in the planning and implementation of their health care. Primary Health Care necessarily implies a cooperation between traditional and modern medicine. Not enough is known, however, about people's beliefs and apprehension of health and illness. There are variations in the types of treatment sought – whether Western or traditional – for different illnesses and for different age groups and sexes. The formal health system needs to have a genuine interest in understanding peoples' beliefs if it is to earn their confidence and cooperation.

In the modern, formal health-care system, most health workers are women. However, they predominate in the low-status occupations while authority and decision-making rests with men. Because of women's crucial importance for health issues they must be included directly and with full responsibility in the planning and decision-making process as well as the process of implementation. Otherwise Primary Health Care will prove an insufficient means of achieving Health for All.

Women as users of health care

Because of the many health risks in the lives of poor women and because of women's crucial role in relation to the health of their children, their access to primary health care should be as free and easy as possible. This is not only a question of the number of health centres available, but also a question of whether the health services are geared towards women's needs and social norms. This is not always the case.

Many barriers exist to poor women's access to health facilities. Long distances, poor communication facilities and the cost of transportation may all be prohibitive. In the rainy seasons, when morbidity is high, health facilities may be totally inaccessible to patients. The loss of a day's wages or of a day's work in the field and the absence of mother from unattended children and necessary household duties are other obstacles. Long waiting hours at the clinic aggravates these problems. During peak agricultural periods there is a fall in the use of services and ill children are more likely to be neglected. Given the fact that nutritional status is low before the harvest season, such a period of illness can prove to be a severe setback.

Shortages of drug supplies in many health facilities means that patients are given prescriptions and have to buy their medicines on the private market. Sometimes the price is totally beyond the reach of the patient; in other cases debts to moneylenders may be the longlasting consequence.

In some countries a lack of female staff at health clinics deters women from using them. Similarly, cultural traditions in some places require that women be escorted by a man when leaving their village. This means that another day's work or wages may be lost, or that women must have the assent of a man before they can obtain health care. Sometimes women are treated with disrespect by the staff.

Younger women look to their elders for advice; if humiliating treatment by health staff is a common experience, women will not encourage each other to go for prenatal care, even if it is available. And when labour starts, a woman will turn most readily to the traditional midwife, whose face is familiar and comforting but who may use unhygienic methods that will pose a threat to her health.[37]

In order to include women fully as a target group in primary health care – in terms of their total scope of life and not only with regard to reproductive problems – health staff at all levels need to be sensitized to perceiving the role and status of women in any given area. Health programmes are too often conceived, implemented and evaluated by specialists without directly involving the recipients of the system. Women are often the focal point and hinge, upon which success of the programme hangs, yet their insight is not used at all. They have been treated as passive recipients of a service system, which may not meet their most urgent needs and which may even be contrary to needs never made explicit.

RECOMMENDATIONS

- The established system of health-care facilities should be decentralized in order to bring services to the people. In public budgets priority should be given to primary health care and essential drugs rather than to high-technology medical treatment.
- Preventive health care should be stressed through comprehensive measures affecting several factors simultaneously in order to break the vicious circle of carriers of ill health. For example, immunization programmes should be paired with health and nutritional education, among other things.
- Health-related statistics should be disaggregated by gender. Services can only be effective when they are based, first, on an accurate picture of health problems and needs of people, and second, on an appreciation of the complex social, cultural and economic factors that affect use of the health facilities.
- Nutrition policy and food policy should be dealt with together at national and international levels. Agricultural planning programmes should encourage the expansion of crops favoured by the poor and grown under the conditions they face.
- Water supply and sanitation programmes should involve women in their design and implementation, since the women have the greatest incentive to make these programmes succeed.
- Working conditions in urban and rural areas have an enormous impact on people's health. More and better attention to worker health and safety will function in many areas as preventive health measures.
- The skills of traditional birth attendants should be raised through training programmes; at the same time efforts should be made to raise their status in the community, so their roles can be broadened and their knowledge shared more effectively.
- The syndrome of son preference/daughter neglect should be known by planners in the health sector. The high mortality rates of girls should be seen as a signal of marginal life conditions, where girls are not taken care of sufficiently.

- The notion that people are poor because they have many children should be questioned. Studies show that when the economic and social conditions improve, the birth rate decreases. Specific family-control programmes have not shown any continued effect.
- The concept of family control should be banished and the concept of family planning be substituted, with birth spacing, which is acceptable to women. In order to be effective, efforts to reduce birth rate should ally themselves with women's wishes and not counter them.
- Family-planning services should be aimed at men as well as women. Condoms, the only safe contraceptive, should be cheap and widely available.
- Because no female contraceptive is totally safe, family-planning programmes should be executed where other health services are provided and women can be given proper information about the risks, instructions on the use of the various devices and follow-up control and treatment.
- It should be realized that the single most influential factor in birth rate is education. The more education provided for girls, the fewer children they tend to have.
- Because of women's crucial importance for health issues, they should be included directly and with full responsibility in the planning, decision-making and implementation of health-care systems.
- Too often women are regarded as objects, not as active participants. Even though they may be illiterate, they should be recognized in their own right as potential carriers of a wisdom which the specialist needs – a vital, valid and down-to-earth knowledge of their own reality, deeply rooted in their own culture and in their own life experiences.
- In most developing countries, good baseline data are not available to health planners. Yet services can only be truly effective when they are based, first, on an accurate picture of the health problems and needs of people, and second, on an appreciation of the complex social, cultural and economic factors that affect the use of health facilities.

NOTES

1 WHO (1978).
2 World Bank (1989).

3 Østergaard (1987).
4 World Bank (1989).
5 McLean (1987).
6 Østergaard (1986).
7 Chauhan and Gopalakrishnan (1983).
8 Mies (1981).
9 UNICEF (1982).
10 Rogers (1980).
11 SNDT University (1983).
12 Royston and Armstrong (1989).
13 Jørgensen (1983).
14 WHO/UNICEF (1986).
15 Moore (1973).
16 WHO/UNICEF (1986).
17 Royston and Armstrong (1989).
18 Levison (1974).
19 Levison (1974).
20 Royston and Armstrong (1989).
21 Royston and Armstrong (1989).
22 WHO (1986a).
23 WHO (1986a).
24 WHO (1986b).
25 Islam (1985).
26 WHO (1986b).
27 Royston and Armstrong (1989).
28 Harrington (1983).
29 Mitra (1985).
30 Andersen and Staugård (1986).
31 WHO (1986c).
32 Ratcliffe (1983); Warwick (1982).
33 Warwick (1982).
34 Warwick (1982); Silberschmidt (1986).
35 Silberschmidt (1986).
36 WHO (1978).
37 Royston and Armstrong (1989).

Chapter 8

Household resource management

Kate Young

The distribution and management of household resources is the last in the series of processes which determine who actually receives the benefits resulting from a development policy or project.[1] As such it can be a critical factor in determining its success.

If practitioners want to make sure benefits reach all members of households or a specifically targeted group, they need to take into account at the planning stage the structure of households operating in the region and their methods of managing and distributing household resources. For example, the Mwea resettlement project in Kenya ran into trouble when the beneficiaries' wives refused to continue working on the project's improved rice fields: the wives were not receiving any reward for their labour, yet were still expected to provide the usual household inputs, like firewood, which they had to buy rather than collect.

The general lack of knowledge and understanding of the dynamics of household resource management is causing increasing concern among development practitioners. This has arisen partly because of the failure of household-focused policies to deliver the intended benefits to all household members, in particular to women and children. A second cause for their concern is the evidence that women are chronically under-resourced relative to men in many poor households. Studies show that even within better-off households, daughters are often physically impaired because they receive a smaller share of resources than sons. Third and more recently, the interest in entrepreneurship has focused attention on the barriers to women's entrepreneurship created by women's lack of control over household resources.

These findings have raised questions about the applicability of conventional models of how households operate as economic units. As the number of studies on the subject increases, we are discovering that our traditional concepts of how households function in the First World no longer fit the facts for our own societies, let alone those of developing countries.

In this chapter we look at what is known about household resource management in both fully capitalist and transitional economies. A case study illustrates the problems that can arise in development projects when intrahousehold distribution is not taken into account. The discussion that follows explains how gender awareness in development thinking can help to avoid those problems. It highlights a number of policy areas and points out some of the policy changes required in order to incorporate a gender perspective.

THE CONCEPT OF THE HOUSEHOLD

One hurdle facing development practitioners who choose to consider intrahousehold distribution in their policies and planning is the theoretical framework available to them. Even though empirically the household is recognized to be composed of a number of different individuals of different ages and sexes, in economics it is treated as a corporate unit, as if it were an individual. Terms such as *household decision-making, household behaviour,* and *household utility* illustrate this. There is little inducement to question how these 'corporate' decisions are reached. In fact a common definition of 'household' is: a group of related individuals who share a home, share meals and who pool their resources for the benefit of the group. If conflicts of interest should arise, it is assumed that the male head of household, in his role of major breadwinner, will act as arbiter.

A second set of difficulties arises from the fact that we tend to conflate or fuse two separate concepts: that of the *household* – a residential unit – with that of the *family* – a social unit based on kinship, marriage and parenthood. As a result many of the assumptions made about the nature of intrahousehold relations are informed by assumptions about the marriage relationship and relations between parents and children, particularly those between mother and child.

In much policy-making, marriage is conceived as a partnership between two persons with reciprocal but not equal rights and obligations to each other, with similar rights over and obligations to the children of the marriage and with a common set of interests. As a result it is assumed that marriage gives rise to a unit marked by joint control and management of resources, and similarity or coincidence of interests and goals. The economic form that rights and obligations are assumed to take within marriage is that of *sharing*.

There are a number of difficulties with these common perceptions. One is the wide range of cultural differences – even in the First World – in the degree of separation of couples from their kin, in the degree to which marriage establishes a joint economic entity, in the coincidence of interests, in the rights to property, etc. Even within Europe, marriage may not be 'joint' in the sense indicated above, but involve a division of responsibilities, separate budgets and no concept of marital property.

Another difficulty is that an ideology of sharing does not necessarily indicate that *equal* distribution of resources to all actually occurs. For example, many studies in the First and the Third Worlds have shown that in food distribution there is a bias in favour of men, whether young or old. The same is found for expenditures on health, education, work tools and personal consumption.

The lesson to be drawn from this evidence is that we cannot assume that familial and marital relations are egalitarian. Rather, we must ask, to what extent do they involve different and possibly unequal obligations and rights, differential power and control?

The household concept and policies for poverty alleviation

Concern with poverty alleviation in capitalist market economies focuses on the wage–price–welfare relation: the household is assumed to be the basic economic unit and unit of distribution, and household well-being is seen to be a function of the relation of prices to wages. Policies to relieve the poverty of identified groups focus on adjustments to the level of income poor households receive – either directly or through state mechanisms. Policies may take the form of raising the basic wages for the jobs that the poor tend to have, of providing a basic minimum wage, of giving tax concessions to the low-paid or allowances to compensate for low wages, etc. Most policies are targeted on the 'man of the family' as the household head.

Serious questions are being raised about the assumptions regarding households which are embedded in these poverty alleviation policies. Four key assumptions are made:

1 That the poor gain their basic livelihood through the money wage form
2 That the household is composed of a couple and their children

3 That the main wage earner or 'breadwinner' is the male head of the household
4 That once the wage enters the household it is distributed to meet the needs of all its members.

In developing countries, the wage–price–welfare relation can be used to characterize the welfare dynamic of only that small proportion of the population who are fully engaged in the wage-labour sector. For the rest, the emphasis must be centred on prices, since in most countries the bulk of the population are either subsistence or commercial farmers or self-employed entrepreneurs. 'Getting the prices right' is the formula for success.

In this model the assumptions made about the form of livelihood are somewhat different from those in the wage–price–welfare model, but the assumptions about the nature of the household and its principal economic agent are similar. Yet as the discussion in this chapter will show, these assumptions about the household are frequently not valid in the First World and rarely valid in the Third World.

Households vary greatly in structure and composition – both within and between societies. Moreover, within a lifetime an individual may belong to a number of differing households because of life-cycle changes. In our discussion we will use the term *household* to mean a system of resource allocation between individuals who may or may not be permanently co-resident, 'in which individual members share some goals, benefits and resources, are independent on some, and in conflict on others'.[2] However, the focus on the household should not be allowed to obscure the fact that individual household members also belong to other groups to whom they have obligations (e.g., a husband to his mother or sisters), from whom they may receive resources and through whom they may gain access to resources.

STUDIES ON INTRAHOUSEHOLD DISTRIBUTION IN CAPITALIST AND TRANSITIONAL ECONOMIES

In the literature, the problem of household resource management is discussed in relation to three sets of issues. The first centres on how to alleviate poverty and in particular how to direct support to 'vulnerable groups': children, older people, female-headed households, the landless or other particularly deprived groups. Methods are sought to ensure that children benefit directly from poverty alleviation measures and that women within households are not made worse off by measures directed to raising 'household' productivity.

The second set of issues is concerned with how to enhance the capacity of the poor to contribute directly to their own welfare and indirectly to that of their country. Since it has been recognized that women predominate among the poorest, many policy-makers have sought to understand the nature of household and family dynamics because this may help to explain how women's poverty comes about.

The third set of issues focuses on how to enhance women's economic contribution to society and in particular to strengthen their capacity to provide for their children. It is now commonly believed that women spend a higher proportion of their resources on children than men do. Therefore, in order to ensure a higher health and educational level for the next generation (to raise human capital endowment), understanding women's relationship to household resources becomes an important development issue. It also raises the question of whether children benefit more if support is given to the father or the mother and whether the benefit should be in cash or in some other form.

Such concerns have led some researchers to examine control and decision-making within the household. The next sections deal with data from both wage-based and other economies and some of the policy implications.

Household resources, decision-making and control in wage-based economies

The common belief is that intrahousehold resource allocation involves income pooling and sharing among all household members; but in fact a number of distinct allocative systems have been identified, each associated with a somewhat different set of gender relations.[3] Jan Pahl suggests that allocative systems can best be analysed by looking at who has the *control, management* and *budgeting* functions, since they represent different points in the flow of money and other resources through the household.[4]

Control is mainly exercised at the point where money enters the household. It is concerned with decisions about which allocative system should be adopted, which spouse should have the final say on major financial decisions and the extent to which each spouse has control over personal spending money and access to joint money. *Management* is concerned with putting into operation the allocative system adopted. This function can extend over all the household expenditure categories, or be confined to just one or two. The criteria that distinguish one allocative system from another are essentially related to management. *Budgeting* is concerned with spending within each expenditure category.[5]

In her study of money allocation within UK households and its relationship to power relations between the genders, Pahl identified four types of income-allocation system: the whole wage, the allowance, the shared management and the independent management system.

In the *whole-wage system* the 'breadwinner' gives all earnings to the household manager and is given back a proportion for personal use, while the rest is used to meet collective expenses. The whole-wage system is usually associated with a non-earning wife, but is also common in a strongly corporate family system with several earners. Monies coming into the household are divided into two basic categories: that used for collective expenditures and that for personal spending.

Where the household manager is the wife, she may have limited control in terms of what to buy, but will have great responsibility to manage the money and usually no right to personal spending money, although she may put aside a few pennies 'against a rainy day'. Where the manager is the household patriarch, his right to personal spending money is rarely questioned.

The *allowance system* is often perceived as the form adopted by households where wives do not earn, but it is also common in those in which they do. This system has a number of variants. In one, the husband's income is divided in three parts to cover (i) basic and recurrent bills (rent, utilities, school fees, medical and other insurances, etc.) which he pays directly, (ii) food and weekly expenses which he pays in the form of an allowance to the wife (called 'housekeeping'), and (iii) his own personal expenses. The allowance may be paid daily, once a week, or over some intermediate period. This form of income management obliges the non-earning wife to ask the husband for personal spending money or money to cover 'bulky' items for the children whenever needed, or to try to save on the housekeeping to acquire a small fund of spending money.[6] In another variant, the husband gives his wife sufficient for her to cover recurrent bills and weekly expenses. The rest he keeps for himself; she still gets nothing for personal spending money. Both variants represent a direct form of control over women's decision-making and even management and involve no concept of a wife's right to separate personal spending money.

In a third variant of the allowance system, which operates in cases where both husband and wife are income earners, the husband may still give the wife an allowance for recurrent and weekly items, keeping the rest for personal spending. Her income is set aside for 'luxuries', bulky items (holidays, extra clothes for the children, school-related expenses), and personal spending money. In fact, much of the wife's income goes to the collective aspect of household consumption and on essential but 'bulky' items, such as children's clothes and shoes.

In the *shared management* system of income allocation, all earners hold aside what they need for work-related expenses plus an agreed amount for personal spending money; the rest defrays recurrent and weekly collective expenditures. In a frequent variant of this pattern, both partners know how much the other earns but keep their incomes separate. They agree on which items of collective expenditure their separate incomes will cover and what each has to set aside for savings.

Independent management is a similar system. In this, husband and wife do not inform each other of their actual incomes, but agree upon a pattern of expenditure which covers all the regular collective outgoings. Whatever is left over remains in the control of the earner. Both this and the shared management system are almost entirely associated with wage- or income-earning wives.

The ideology of sharing, companionate marriage and meeting household members' needs whether or not they are contributors may be declared whatever the actual system used. But the typical pattern of outlays seems to be that a higher proportion of women's income, whether earned personally or given as an allowance, goes towards the non-durable expenditures associated with the collective aspects of family life – and particularly on the children – while men's earnings go towards consumer and other durables as well as personal spending.

Research has uncovered some interesting perceptions about income 'ownership' despite declared ideology of sharing. For example, in the United Kingdom, state benefits are more likely to be given to the wife to manage because they 'belong' to the family while wages 'belong' to the earner. Men's basic pay is earmarked for collective expenditure ('ours'), while overtime or bonus pay is for personal spending ('mine'). A similar distinction is made between salary and additional earnings.

These perceptions may affect men's take-up of work opportunities. For example, men who turn over their whole wage to their wives seem to do less overtime than those who give their wives an allowance. For similar reasons, men may prefer to have a rise in bonus payments rather than basic pay. When men have two jobs – an official and an unofficial one – they may consider only the 'official' earnings as being for collective consumption. Pahl suggests that the growth in the informal economy may in part be a product of this division of earnings.[7] Women also have unofficial, occasional, or seasonal earnings. However, research indicates that these are usually tied to specific 'big' expenditures: a child's winter coat, school clothing or outings, or even the family's summer holiday.

Decision-making and control

Studies on household resource management have tried to find out what determines the choice of allocative system and the implications of adopting each one. The findings suggest some explanations but are slightly contradictory. Pahl (1983) points to three main variables: the level of income, the source of income and the normative expectations of the local culture.

Level of income

Evidence, largely from northern industrial countries, suggests that the whole-wage system is associated with poverty. Middle-income couples use the whole-wage or the allowance system, while upper-income couples use the allowance, the shared management or the independent management system. Three implications can be drawn from this.

First, that women are given the task of managing poverty and stretching the wage. They invest large amounts of unpaid labour in 'shopping around' and providing a whole series of goods and services otherwise procured for a price on the market. Second, that when more substantial resources are available, men take over control and management. Hence the suggestion that 'greater affluence for the couple leads to greater inequality between husband and wife'. Third, that when incomes are sufficient to permit some degree of flexibility, women may be given greater say, particularly if they are wage or income earners themselves.

Source of income

Research to date supports the contention that earning wives have more say in the allocation of household resources than non-earning wives. But it does not follow that the higher the wife's contribution to the total budget, the greater her control or say over how joint income is spent.

The connection between income-earning and power in the household is by no means a simple one. Factors such as the wife's age on marriage, the difference in the ages of husband and wife, the number of children, their ages and each spouse's ownership of other economic resources, including capital and land, also impinge. Extra-domestic factors can be critical, too, such as each partner's possession of social resources; the prestige of occupation, the natal family's social position, the social network, etc.

Normative expectations

The power a wife may wield is also shaped by ideologies about the nature of marriage and by the expectations held by the couple, their social network, local community or even occupational community. Many wives feel a husband should control family affairs, including finances.

Policy implications

As illustrated above, research over the past decade has shown there are errors in many of the assumptions made about patterns of household-income management which are embedded in our policies to alleviate poverty. This suggests that we need to revise certain policies and develop new methods for assessing their impact.

Because of the unequal allocation of money within households, policy-makers need to be able to assess income on an individual as well as a household basis. Assessments of poverty based only on household income may conceal much hidden poverty. For example, studies of divorced women on supplementary benefit in the UK indicate that up to a third of them consider themselves better off on supplementary benefit than when they were married. Since supplementary benefit is pegged roughly at the poverty line, this means the women – and possibly their children – were presumably living below that level as wives.

Within social security policy, it is probably more effective to pay benefits directly to women – especially if children's welfare is the main concern – than to augment subsidies to men. This is true even when women are in paid work, because the woman's income goes largely to the collective aspect of family expenditures.

In wages and salaries policies, to alleviate want within families, a strategy to increase basic pay is more likely to benefit women and children than one to increase bonus payments. This goes against the grain of much present thinking on incentives, but policy-makers need to recognize that bonuses and overtime pay are often seen as personal rather than collective money.

Taxation policy should also be reviewed in light of the finding that income is not evenly distributed within the household. A shift in policy from direct taxation and state subventions to lower taxes and subventions with indirect taxation may be harmful to non-earning household members. Those whose incomes increase may not automatically transfer their additional income to those who have to cope with the higher prices.

HOUSEHOLD RESOURCE MANAGEMENT IN TRANSITIONAL ECONOMIES

When discussing household budgets in developing countries in terms of models developed in the industrial market economies, a number of problems immediately arise:

- Household income is likely to be made up of a wider range of items than wages or income from investments
- The composition of domestic groups is more varied and so are the patterns of authority within them
- Marriage may not be joint
- The resources of a couple may be managed by others within the same household or spread over a number of other units
- Cooperation *between* households may be just as significant as collaboration *within* a household in assuring the members' ability to meet basic necessities.

Types and sources of income

It is important to recognize that developing countries differ considerably in the degree of dominance of the market throughout the economy. In some – like Ghana, Peru and Burma – a sizeable minority of the population is not fully integrated into the market. In other countries, such as Chad, Bolivia and Nepal, the bulk of the population is not. Even within a single country there are wide differences in the degree to which sectors of the population are incorporated into the market. This means that households are not structurally homogeneous: while some of them are largely dependent on the market for survival, many are not.

In many developing countries, the majority of households, whatever their income source, live at or below the poverty line. The resources which these poor households have are likely to vary in terms of the *nature* of the resources:

- Market or subsistence goods acquired through production, barter, exchange, state provision, remittances, or on credit
- Services owed
- Cash from paid work, market sales, interest on loans, hiring-out of tools, remittances, state provision, etc.

The resources will also vary by season and over time, as children become productive and as adults give up some types of work. Thus people gain access to the necessities of life primarily by two means: work and the 'support system'.[8]

Access via work

Work can be paid either in cash or in kind. Its productivity depends on the capital (land, tools, equipment, animals) and the technology that a worker has at his or her disposal, as well as the worker's *human capital* assets (health and education). Work is also influenced by the sexual division of labour, which imposes different constraints according to the country and culture.

In agriculture, for example, in some countries women are virtually full-time farmers, while in others they do little work in the fields. Where women are active farmers, in some places they work side-by-side with the men in the fields and in others there is a separation between men's and women's activities.

The commonplace belief that men are involved in cash crops and women in food-crop production is not supported by research. In many cases, building this idea into internationally funded development projects has resulted in negative consequences both for the women involved and the projects themselves.[9] In fact, men grow food crops in many areas and in most regions women grow crops for the market as well as for the home.Yet it is true that almost everywhere women do seem to predominate in processing raw materials either for family consumption or for the market.

The reasons why the sexual division of labour takes the form it does in most countries are not entirely clear. Whitehead suggests that women are more strongly associated ideologically with the collective aspects of consumption than men.[10] The First World research findings discussed above suggest a similar association.

The sexual division of labour appears to be more marked in the wage-labour sector, whether in agriculture or other forms of production. This means that different income items are likely to come from the specific labours of men, women, boys and girls. Ensuring adequate consumption for a poor household then depends on:

- Having the right mix of labour available in the household, or being able to acquire it through kinship or other support networks
- The nature of production cycles and the type of employment available
- The ease of access to loans (especially in kind) and credit.

For many households, acquiring cash may be the most problematic aspect of income, even though the amount they require may be extremely small. This is a recognized fact in rural areas, but even in urban and peri-urban slums there are large numbers of households who appear to survive on a virtually cashless budget. In this situation, the person who is able to get cash on a fairly regular basis may acquire a more dominant position within the family, at least ideologically.

Access via the support systems

Public support systems can offer social security, welfare, school lunches, free medical care, etc., but such state aid is not well developed in the Third World. Kinship remains the primary support system, supplemented by friends and/or other mutual-help networks. The support these private systems offer can take the form of transfers of money, goods, labour or other services. Sometimes the transfers occur regularly, e.g. when migrants remit money, when a daughter supports her siblings through school, or when a man provides regular support for a widowed or separated sister. Sometimes they occur irregularly, as gifts for special occasions or when the donor has a surplus.[11]

Despite the diversity of kinship systems, in general they structure access to and control over property and other resources rather differently for men than for women. Women's access to basic productive resources is largely indirect and often by means of a relationship with individual males: husband, father or possibly brother. A successful claim may depend on the quality of a woman's relationship with a particular individual, who may have different interests.

As the economies of most developing countries have been undergoing change, traditional rights of access to productive resources have been coming under strain. Much of the development literature points to changing practices around claims to resources or to right of support. For example, widows have increasing difficulty in pressing claims for economic support from husbands' kin, as do divorced women from their own kin. This decline in support is partly caused by the fragmentation of kin groups (due to growing differences in wealth and class) and by the poverty of the majority of the population.

The commoditization process also undermines communal forms of property and erodes the resource base of kin groups such as lineages. As the market in land develops, land once held communally is fragmented into individually owned plots; kin groups such as lineages are no longer able to allocate use rights on land to new lineage members. As this happens women's claims on kin resources appear to erode more swiftly than men's. Below we give some examples concerning land, housing and capital.

Land

Even where both men and women can claim rights in land, the scope of these may differ. For example, in much of patrilineal sub-Saharan Africa, a woman's access to land is by virtue of her wifely status, while her husband's is by virtue of his lineage membership. A wife is given access to basic productive resources of the husbands' kin group inasmuch as she will become a mother and have the responsibility for feeding the future generation of the group. If she has the misfortune not to have children (or even sons), she stands to lose her rights or to be given extremely marginal land, while her old plot may be given to her husband's next wife.

Even if a woman bears children, her use rights are often to land in general and not to particular plots of land. As a result, she runs the risk that having improved the productivity of her allocated plot, another poorer plot will be substituted for it. She is unable to prevent this.

Muslim law offers another type of example. Women do inherit rights in land, but it is their husband who will manage it and, as such, may dispose of it. This underlines the important distinction between ownership and possession. Given this, a common strategy for women is to cede their land rights to brothers in the hope of support should their marriage fail. A common strategy of fathers is to marry daughters to close patrilineal kin, thereby keeping them and the land within the lineage.

Housing

Little direct empirical research has been carried out on housing, but in a number of countries it appears that divorce or widowhood frequently involves women losing all rights to their marital home while marriage gives them weaker rights to their natal home.

In India, recent research in Calcutta shows that widows find it difficult to stay in the home of their deceased husband, while unmarried women often find it impossible to remain in their natal home (especially if their married brother is living there).

In much of patrilineal sub-Saharan Africa, a widow has no right to the conjugal home and upon her husband's death its contents may be divided among her dead husband's kin. In the past one of the husband's close male kin would have inherited responsibility for the widow. But this practice is now meeting with resistance from both widows and the husbands' kin: they are reluctant to take on the economic burden of an elderly woman, particularly if she has no sons.

Capital

Studies of women entrepreneurs show that acquiring start-up capital is often very difficult. In sub-Saharan Africa, where in-marrying wives are seen as a reproductive resource for the man's lineage, women looking for start-up funds for personal enterprises may have little claim on their husbands' resources. Research among women traders shows that such capital is acquired from female kin (mothers and aunts) as much as from husbands and that usually the amounts are extremely small.

In systems of pre-mortem inheritance, upon marriage daughters get their share of the family's wealth which exists in movable form – jewellery, cash or consumer durables. But this may be treated as the common property of the couple, or may even be managed by the husband so a married woman may be unable to control its further transmission. This is argued to be the case with married women's dowries in urban India.

Women's access to the wider resources of society through credit is also circumscribed. Even when they have some assets – such as jewellery, livestock or land in usufruct – these are often not recognized as adequate collateral.

Thus, in general in developing countries women face great difficulties in obtaining or being endowed with valued resources which they can use to build on or convert into other assets. They also find it difficult to hold on to the assets they have in the face of claims by male kin or spouses. Although many countries have modified their laws concerning married women's share of conjugal property over the past twenty years and have passed new laws permitting women's ownership of property and access to credit, discriminatory notions are enshrined in common practice. This is proving far harder to alter.

Marriage and the household

Any work on household resource management should involve an understanding of what marriage involves in a particular culture, especially the degree of 'jointness' of a couple and their incomes and assets. There is a very wide range of different marriage and residential arrangements. In addition to the nuclear household, for example, there are residential arrangements where:

- A man and his wives co-reside in his compound
- The couple lives with the husband's parents, brothers and their wives
- The husband lives with his mother and sisters and only visits his wife
- The husband moves in with his wife and her parents
- The household comprises a woman, her children and her mother, with the father(s) of the children paying occasional or regular visits.

Thus, a woman may marry into a residential group of which her husband's father is the head and in which his mother manages household resources. She may marry a man who already has a number of other wives. She may marry but remain in her natal home. As a result of these different marital and residential patterns, the ways in which resources are managed varies widely. In many of these situations, the notion of economic jointness is absent.

Household vs. individual resources

The ideological separation of men and women within marriage has interesting effects on spending patterns. In a number of West African societies, for example, both husband and wife have their own resources and maintain separate budgets. Men's and women's incomes are rarely allocated to the same expenditure categories. Instead, cultural traditions broadly determine which aspects of collective expenditure each must cover, supplemented by agreements between the husband and wife.

A common pattern in urban areas of West Africa is for women's income to be allocated to day-to-day food (which includes feeding *all* kin living in or visiting the household), clothing needs and domestic goods: cooking utensils, basic materials for house maintenance, etc. Men's wages or income is used to cover regular bills (electricity, gas, etc.), rent and/or the cost of buying the house or house plot. Children's schooling may be paid for by either parent, but most often fathers cover fees while mothers cover day-to-day costs like books, pencils, uniforms, etc.

These patterns allow a critical difference to develop between husband and wife over the long term, which has interesting demographic implications. Women's expenditures vary according to the size of the household (the number of children, visiting relatives, foster kin, etc.) and in large measure with the age of the children; i.e., they increase as household size increases until the children become earners or leave the natal home. Men's expenditures remain relatively stable; rents and utilities may increase, but this may be compensated by increased wages. Thus there is a possible divergence in the interests of husband and wife regarding family size.

In India, the concept of women's property seems to be absent. Even the property that the wife brings into the marriage is seen as the property of the marriage and passes into the control of the husband or his kin. It is not culturally appropriate for wives to have a separate income or budget; indeed, many are not permitted to work outside the home. Those who do generally hand over all earnings to the household head or manager for reallocation. In-marrying wives are not seen as co-equals with their husbands but very much as their subordinates; in many cases the wives are also subordinate to all other members of his family, particularly all other males. As the woman grows older and as she bears children – particularly sons – the degree of her subordination decreases. If she ever achieves the status of wife of the patriarch, then as her husband's proxy she probably also simultaneously achieves the status of family-resources manager.

These two examples reflect the findings of research to date in transitional economies. In summary, they lead to the conclusion that the proportion of a husband's income or resources that he spends on his children or his wife is very variable. In certain cultures a man's chief obligation is to his lineage and his male kin; in some matrilineal societies to his sisters and their children and then his children. In some cultures his primary loyalty is to his mother, not his wife.

Given a man's divided loyalties, a husband may allocate his income in a way which does not correspond to the immediate needs of the conjugal unit or his family of procreation. For instance, he may pour his resources into his own kin. If he lives with his wife and children, the impact this has on them will depend on:

- Whether the wife is economically independent
- Whether he can command the resources of his wife and children for his own enterprises
- Whether he can use 'family' resources to cover his personal maintenance costs.

An African man may adopt a strategy that focuses on providing for his children (particularly his sons) but not his wife; she may be expected to make do on the minimum or to be an independent earner. The effect of this system on his wife depends on how much she is expected to provide for the children out of her own earnings and on whether the economy actually provides adequate employment for her.

Wives are more constrained in their income-allocation strategies, partly because of their close ideological association with their children's welfare, particularly feeding.[12] This association appears to be independent of whether or not the mother has rights to her children. It might seem logical that where women are likely to lose their children on divorce they have a lesser commitment and invest less time and effort in them. However, since proper child care is a critical element of the conjugal contract, where women have lesser rights to children they may in fact invest more in them.

In choosing an allocation strategy, a wife's concern may be to divert as much of her income as can be spared from the children to her own kin, or in acquiring property or durables that are clearly understood to be hers. Increasingly in urban sub-Saharan Africa with patrilineal systems of inheritance, to avoid difficulties on widowhood or divorce, a wife who acquires household durables immediately ensures that both sets of kin acknowledge the goods as her property. If a wife is not permitted to earn an independent income (as in India), her strategy may focus on ensuring her sons survive and instilling in them an enduring sense of love, loyalty and obligation to her, rather than to a wife.

Most descriptive accounts of women's expenditures in poor households emphasize the very small quantities of money that pass through their hands and the countless ways they try to expand their income by preparing cooked foods, handicrafts, providing services, buying animals or poultry for fattening, joining informal credit associations, lending at interest, borrowing against future services, etc. Most of this income appears to go towards children or to the family in general.

The literature also emphasizes the small amounts of personal spending mothers allow themselves, even when their needs are pressing, such as for medical care. Men, on the other hand, claim a right to personal spending money whether or not they earn a wage. For example, peasant producers and subsistence farmers almost everywhere have a right to expend family resources on conviviality and status consumption.

Differences in claims on resources

Because of the very wide variety of household types and the varied meanings of marriage, far more research is needed before we can pinpoint the common variables which underlie different patterns of household-resource management at lower income levels. For the present, only a few very general observations can be made.

One is that in many societies the extent of a man's obligations to his wife depends:

1 On her performance of her duties to him – i.e. on whether she is a 'good' wife, or
2 On her performance of her duties to 'his' children – i.e. on whether she is a 'good' mother, or
3 On her performance of her duties to his kin – i.e. on whether she is a 'good' daughter-in-law.

The reverse does not seem to be common. In practice this means that men can exercise greater control over wives' labour ('good' wife) and over their resources ('good' mother) than vice versa. In those cultures where men's first loyalty is customarily to their own kin and wives move into their husbands' natal home, a wife's position may be too weak for her to make claims on so-called conjugal resources, even though in the end the children would benefit. This weakness of wives in relation to their husbands' resources would not be particularly grievous if women had a strong claim on their own natal families or lineages. But in most instances they do not.

Another general observation is that daughters have lesser rights to familial or kin resources than sons, not only to wealth-producing assets (land, etc.) but also to consumption items. One result of this is that, when grown up, the daughters are often at a disadvantage to men in terms of education, skill, training, health and nutritional status and their capacity to build up or even undertake productive enterprises is greatly circumscribed. Another result is that many women start the responsibilities of married life from a markedly weaker social and economic position than their husbands. This clearly affects their initial bargaining position within the marriage.

Women's claims on household resources

Choosing the variables that enable women to have greater control and decision-making in household resource management is a complex task. We have to ask: does the concept of household resources actually exist in the culture in question? Which women should be targeted?

When we discuss people's rights to make unilateral decisions about the disposition of their income, assets or resources, we should not forget or ignore the familial and social sanctions that can be brought to bear on women or men to ensure that the decisions taken are within defined limits. The literature to date appears to show that culturally women are more firmly associated with the collective aspects of household/family consumption, while men are associated with the security or political standing of the family, household and kin group. This permits men to take a more self-centred approach to their own and family resources and to their own consumption.

In this sense there is some similarity between developed and developing countries.[13] In developing countries, independent access to income and resources may help wives to change the terms of the conjugal bargain and provide them with a stronger bargaining position. But this will not necessarily be transformed into greater power and control over budgetary allocations. To understand this type of power within the domestic setting, we first need to analyse and understand the differences in the social position and relative social value of men and women within their kinship group and the wider community.

Social value

The clearest indicator of the difference in the social value of men and women within the household and kinship unit is the differential spending on boys and girls. Most societies exhibit a slight preference for sons; the Hindu and Muslim societies and those influenced by Confucianism show a marked preference. The World Fertility Survey of October 1983 gives figures for 40 developing countries. While the reasons for the preference vary, the effects are depressingly similar.

Boys are breast-fed longer and given more nutritious or greater quantities of food even in those households not suffering the effects of poverty. Thus if we are trying to eliminate the discrimination in food intake of girls (and adult women), increasing the food availability for the family is a necessary but not sufficient condition.

As a consequence of differences in feeding, girls in many societies have a lower nutritional status than boys. Indeed, in India the most significant determinant of nutritional status was found to be sex. Yet when they are ill, girls are given lesser care within the home and less access to modern medicines or treatment. As a result they suffer higher rates of morbidity and in some cases mortality. The long-term effects of such continuous neglect are not fully understood; they undoubtedly influence general physical abilities and are linked to difficulties in pregnancy and childbirth. Undernourished mothers give birth to low-weight babies because of their poor health and so the cycle of deprivation is repeated.[14]

Consistently lower levels of spending on education for sons and daughters is another indicator. This is clear from the statistics on educational levels, take-up of vocational training, etc., although there have been changes in some regions in the last decade.

What are the reasons for this pattern? Since mothers are charged with feeding children, why do they feed daughters less? Since mothers are usually charged with the health care of children, why do they cosset daughters less?

In societies where women's rights to kin or conjugal assets depends on bearing at least one son, and where women's comfort (even survival) in older age depends upon filial care, differential feeding and care of sons might be a necessary strategy for women in poor families. But these differences are found even in families who are not short of resources. This indicates that mothers are either not aware of their behaviour, or that it is justified in ideological terms.

Many of those working with women's development are finding that women's low self-esteem acts as a barrier to improving their health, education, income-earning capacity, etc. For instance, those involved with health issues report that women's lack of recognition of their right to health results in their not knowing or ignoring symptoms of illness. Thus one hypothesis for mothers' neglect of their daughters is that a person who has so little self-esteem that she can deny the illness of her own body is likely to be profoundly alienated from the physical, psychological and emotional needs of others like herself. A more utilitarian suggestion is that mothers are merely preparing their daughters for the harsh realities of their social fate.

The lack of women's social worth also has important implications for the nature of relations between men and women in society. These relations are traditionally seen as complementary. The family in particular is seen as an unproblematic arena in which men and women collaborate in terms of a set pattern of sex roles derived from (unchanging) human nature. But recent research has found that the family appears to be one of the major battlegrounds of the sexes. Men and women struggle with each other to have their own needs met and to further their own advancement – often while also trying to further the collective good. A number of analysts now look at household dynamics in terms of bargaining theory.

Amarta Sen suggests that the household is most usefully represented as a case of cooperative conflict.15 When spouses have different goals and strategies, there are a number of potential solutions which Sen terms 'collusive agreements'. The one finally adopted is the result of the bargaining ability of the couple. Ability in this sense, however, is not a personal negotiating skill, because the spouses do not come to the bargaining table with equal power.

The perception of self-worth is one important component in bargaining power. This differs widely between family members, thus influencing perceptions of advantage. It is based in part on the 'perceived contribution' that the various members make to household well-being. Two factors enter here: the actual ability to earn income or to bring valued resources into the household and the value given to that contribution by other household members. Men often get privileged access to income-earning opportunities and women's contributions to both household and external labour are generally underrated; thus, men's already strong position is reinforced. The person perceived as providing the main inputs to the family or society is given the further right to decide about resource allocation. Women with a low sense of self-worth have even weaker bargaining and fall-back positions.

Sen's concept of cooperative conflict nicely captures what we suggest is a realistic interpretation of human behaviour: the mixture of cooperation and competition, of seeking own aims and furthering those of others. His analysis points to the need to be aware of the factors that turn a balanced relation into one of inequality. Sen argues that a number of forces have conspired against women's bargaining position over time and that the asymmetries that develop are sustained. Thus the relative weakness of women in cooperative conflict in one period tends to remain stable and sustain their relative weakness in the next period.[16] Until women's contributions are recognized and made visible, therefore, their bargaining power will remain weak. And this indicates that one of the most important areas of policy must be public education to reverse the ideologies of gender inequality.

SEX ROLES IN THE NIGERIAN TIV FARM HOUSEHOLD: A CASE STUDY

Development policy-makers and practitioners assume that differences in the allocation of household resources by gender have little impact on agricultural productivity. However, this assumption is now being challenged.

In 1985, Kumarian Press published a series of case studies for planners which demonstrated that development efforts have different impacts on men and women and on different classes. One of the studies, by M. Burfisher and N. Horenstein, showed how gender analysis can be used to re-evaluate and improve *existing* projects and programmes, as well as new ones.

They looked at a major project aimed at improving agricultural productivity and increasing farm incomes in the 'middle belt' of Nigeria. It includes a basic technology package of improved inputs and new/improved cultivation methods, training and extension programmes, commercial services and water, road, livestock, fisheries and forestry development. The original project design used the household as the basis of analysis, took as givens aggregated labour use and income data and assumed that decisions about the allocation of household resources were made by 'the household'.

In re-evaluating the design of the project while it was being implemented, Burfisher and Horenstein used data from the project's planning documents juxtaposed with ethnographic material on sex role differences amongst the Tiv. They then assessed the abilities and incentives of each sex to adopt the technologies being introduced. They were thus able to estimate the project's likely impact on each sex, relate this to the real results and evaluate the probable outcome of the project had the gender division of labour, income and financial obligation in the Tiv society been taken into account at the start.

There are four reasons why any efforts to increase farm production and productivity in sub-Saharan Africa need to give explicit attention to both male and female farmers:

1 Labour availability at critical times of the year is a chief constraint on agricultural production in the region. This affects the ability of farms

to use more than a certain amount of land and to adopt new labour-increasing technologies.

2 There is a sharp division of labour in the average household: typically, women and men control different crops, carry out different tasks and spend very different amounts of time on farm and household labour.
3 Women provide up to 70 per cent of the region's domestic food supply and perform 60 per cent to 80 per cent of all agricultural work. In addition to food production, women play major roles in food processing, petty trading and farm labour.
4 Male and female farmers have different sources of assets and incomes and different financial responsibilities. Women are responsible for food and for clothing themselves and their children. Men are responsible for large farming and family expenditures and personal expenses.

The Tiv farm and environment

The Tiv farm the savannah lands to the north and south of the Benue River in central Nigeria. For the most part they are subsistence farmers and farm resources are limited to labour and land. Few farmers have access to the limited commercial credit that is available; most depend on local or traditional sources for their seasonal needs. Many women and men belong to traditional associations. For women, these often take the form of contribution clubs which aim to assist their members in small-scale capital formation.

Women and men both play major roles in crop-production activities, but their roles are sex-segmented and sharply differentiated. Work on the farms' staple crops is done by both genders, but each performs different and generally complementary tasks. Very few are performed by both women and men.

Children assist parents in farm tasks and carry much of the responsibility for early child care, freeing the adult women for other kinds of work. Thus, child labour is an important component in the Tiv's farm-labour resources, but government policies for universal primary education could disrupt this balance.

Other income-generating activities include the processing and trading of agricultural products, spinning and weaving and pottery-making. Trading is considered a woman's activity, so women sell not only their own crops but also a portion of the output from men's fields.

Every married Tiv woman has a right to a farm of sufficient size to feed her own family; this is allocated to her by her husband. Unmarried women have no land rights and widows continue to have land rights only if they remain resident within their husband's compound. The women determine the use made of their land, own the produce grown on it and decide how much of the crop should be given for domestic consumption and how much for the market.

As Tiv women are responsible for food production, they control the major portion of the Tiv's subsistence foods – yam and sorghum. The women get income from selling surplus basic staples, and from trading in a large variety of side crops or in food prepared from these crops. They use this income to provide for themselves and their families in accordance with their obligations.

Men have specific duties towards their wives. These include allocating fields, clearing land, hoeing yam mounds and providing seed yams for each wife's first yam crop. Men control the incomes earned from the crops they grow: millet, rice and some cassava. They also earn income from weaving and fishing and contribute to the family food supply through their hunting.

The Tiv's system of intrahousehold exchange reflects their values of the collective responsibility to share as well as to give payment in kind for the performance of specific tasks. Therefore men frequently give millet to women to process and sell on their own account in return for women's labour on the men's fields and vice versa. In certain cases foodstuffs may be sold within the household. For example, women will buy millet from their husbands to make beer to sell, or sell their husband's crop and keep the profit. Spouses often make loans to each other, frequently with interest attached.

Women do not participate in the local political system; they are represented by their husbands in all matters of importance concerning the family. However, older married women can exercise informal power within the compound because they have control over the food supply and also hold a great deal of authority in domestic affairs.

Re-evaluation of the project plan

The project planners estimated that on a hypothetical 2.5 hectare farm, total labour requirements would increase by 14 per cent annually. They saw this as an obvious constraint and planned for potential bottlenecks by encouraging a slight decrease in early season crop planting.

However, when Burfisher and Horenstein analysed the labour

requirements disaggregated by sex, they found that women would bear a disproportionate share of the increase. Because the project planned to increase production by improving yields rather than expanding acreage, most of the additional work would come in harvest and post-harvest activities, where women have primary responsibility.

The project planners estimated that the net returns to labour would increase by 31 per cent and that women's incomes would rise slightly more than men's. Moreover, since food crops were expected to improve their yields most substantially, women would benefit more from the production increases than men. However, Burfisher and Horenstein found that the net returns from women's crops would be mostly in kind and the actual cash component small, while men's crops would have a greater cash component because they are mainly marketed.

The re-evaluation found that the non-corporate household had several implications for the project's outcome:

1 The introduction and adoption of new technologies could not depend on pooled family labour as a resource or on shared income as an incentive.
2 The new technology being introduced was not in the areas of harvesting and processing, where women have primary or sole responsibility, yet women's labour was being increased disproportionately to men's. These factors could impair the women's ability to meet new labour requirements and reduce their productivity relative to men's.
3 The project did not address many labour-intensive activities of the farm household, such as fetching water, cooking, food processing and the harvest of women's side crops. Time spent by women on these activities would inhibit them from allocating more time to the crops designated by the project.
4 Both sexes had potential to increase income, but increased labour requirements were not always associated with increased income.
5 Non-financial considerations, such as women's responsibility for family nutrition, could determine if and to what extent new technologies were adopted by farm-household members.

Recommendations

In general, the different roles, constraints and incentives of each gender can cause projects to have unintended effects and problems which are not foreseen by conventional analyses based on the corporate household and aggregated data. Burfisher and Horenstein gave five general guidelines to help project planners take gender differences into account.

1 Socio-economic baseline data should be sex-disaggregated at the project design stage. The household should be viewed as an integrated production and consumption unit, with all relevant activities of household members taken into account.
2 Any increased labour component for women caused by the project should be counterbalanced by the timely introduction of labour-saving storage and processing techniques. This could not only save labour time, but also provide women with a source of cash income.
3 The shift of child labour from housework and other activities to schooling may have an effect on women's ability to participate in the project and should be analysed.
4 All training and extension courses should be targeted at both women and men, taking their needs into account. Provisions should also be made to ensure the women can participate in the courses.
5 If women lack representation in a local political structure, they should be ensured adequate access and representation to project personnel.

The Kumarian Press series of case studies are highly recommended. Not only do they show how to take account of sex-role differences when planning projects, but in Burfisher and Horenstein's study the differences in impacts are quantified.

FUTURE RESEARCH QUESTIONS

If development policies, programmes or projects are to be effective in raising the standard of living of the poor, understanding how households function should be an essential aspect of basic planning knowledge. We need to ask under what conditions husbands and wives pool their resources or allow each other to have access to separately acquired resources?

- Are the conditions purely cultural – patrilineal kinship, polygamy, co-residence of parents and sons, etc.?
- Are they economic – associated with the main way in which livelihood is acquired?
- Does jointness of income management benefit all family members more than separate management or the allowance system?

- Are there any positive benefits of the allowance system for women or children?
- Under what conditions do men withdraw or decrease their financial support for their family of procreation? Can systematic patterns be found?

It is important to find answers to these questions because a great deal of national policy-making and international project development is based on assumptions about the sharing and pooling of couples' incomes.

POLICY IMPLICATIONS

Understanding the real patterns of income, spending and use of household and family resources among the urban and rural poor will have implications for policy-makers and project designers. Here we look at three general policy areas.

Recognizing the need for much more detailed information

We have a rather limited knowledge of the total resource package that makes up the income of poor people. The degree of dependence on cash is not even clear for poor urban families, let alone rural subsistence producers. We do not know to what extent cash is a general household resource or is allocated to particular expenditure categories or particular household members.

We do not know which household members contribute what to the overall budget. In large part this is because our censuses and surveys are designed according to assumptions about how family-based households operate.[17] This may lead to considerable distortions:

- If questions focus on the economic activities of the chief earner of the household (assumed to be male), the economic activities, hours and returns to labour of other household members (women and children) are neglected
- If data about the activities of other household members are obtained from the 'head of household' – given the fact that in some cultures women are not given public recognition as economic actors and that in others men may be genuinely ignorant of the incomes/activities of the women and children they live with (particularly where income is not pooled) – the data collected will be one-sided
- If patterns of income and expenditure fluctuate widely over the

seasonal cycle, surveys and censuses that assume uniformity throughout the year and collect data from a single reference period will inevitably make invisible critical economic processes.

The solution is to find ways to capture data on domestic budgeting arrangements on a regular and thus comparative basis. An essential part of such data collection should be getting details of the proportions of income earned by and spent on each member of the household. The systematic collection of such data among particularly vulnerable groups would make it possible to devise an early warning system.

Greater awareness of the potentially disruptive effects of projects and policies

The knowledge that households have varied patterns of acquiring a livelihood which use all available household labour should make planners and project designers think twice about a number of assumptions they make.

For example, conventional ways of organizing schooling may involve removing crucial income-generators from the family when needed. This can lead to lower family income, or to unmanageable work pressure on one or both parents, or to fluctuating attendance, a high rate of school drop-outs, discrimination against girls' education, etc. An alternative method would ensure that access to learning for child workers is scheduled around their need to be economically active at certain periods of each day.

Another example is labour use in agriculture. Projects supposedly designed to improve general productivity often improve only men's productivity and increase their control over family income and labour. When a project builds on a supposedly freely available supply of 'family' labour (i.e. women's labour) to men's activities, it disrupts women's existing, essential economic activities. Usually it also leaves women struggling with low productivity, labour-intensive methods of cultivation, processing and/or storage. In some cases this may have little impact on the project's own terms of reference and criteria for success; in other cases the project itself fails. As far as policy is concerned, it can be argued that the neglect of women's productive activities in sub-Saharan Africa has contributed to the current food crisis in that continent.

The solution is for project designers and others to assume that *all members* of the family contribute to family welfare, unless they have data showing this is not the case. Planners should also assume that the sexual division of labour allocates different tasks and activities to the genders and that substitutability of labour is problematic. It should be the objective of *all* projects and policies to raise the productivity of *all* household members.

Clearer understanding of the impact of planned interventions

Planners' ignorance of the nature of gender relations in any given society at the categorical, community or household level can lead to disruptions of the balance between the genders, or indeed sharpen inequalities that already exist. It is common to hear disavowals of any intent to interfere in this fundamental human relationship, but the evidence of past decades shows that in many areas both projects and policies have had the effect of shifting the balance of power towards men. The common litany that women have been denied access to information, training, agricultural extension, credit, inputs and access to specialized markets; have lost their traditional access to land, kin and community support systems and other resources; are absent from the arenas where crucial decisions are made that affect their lives as much as men's, testifies to the effects of both planned and unplanned economic and social change.

Blame for the unplanned changes cannot be laid at the door of project designers and policy-makers except in so far as they promote policies that indirectly result in such changes. But they must bear responsibility for many of the *planned* changes. It is no longer credible to plead ignorance given that evidence of the negative effects has been available for at least a decade.[18]

The solution is for development practitioners to make a much more serious attempt to bring a gender focus to their own work. A first step is to look carefully at the literature on gender relations. A second and equally important step is to look at the conventional categories of macro- and micro-economics and to ask whether they too are the product of certain assumptions as to the form that gender relations take. For example, is it useful to focus on the production and distribution of goods and services, but to ignore the production of the human agents that make or distribute them? Is it sensible to discuss technology and other productive inputs without spending time analysing the social arrangements that permit the technologies and the inputs to be used and the productive process to be carried on? What is gained by taking a view of the pattern of activity of households that excludes the work that goes on inside the house – and, indeed, some that goes on outside it, such as fetching wood and water? Is it useful to value the work of the food cultivator but not its processor?

Ensuring that the benefits of development interventions are distributed to all members of households as equitably as possible makes sense once it is recognized that all members of households contribute to the welfare of the unit and to the wider society.

NOTES

1 Rogers (1985).
2 Feldstein (1986).
3 Cf. Dwyer and Bruce (1988).
4 Pahl (1983).
5 Pahl (1983, p. 244).
6 Cf. Roldan, in Dwyer and Bruce (1988).
7 Pahl (1983).
8 Mueller (1983).
9 See Chapter 3: *Agriculture.*
10 Whitehead et al. (1981).
11 Mueller (1983, p. 272).
12 Whitehead et al. (1981).
13 See Whitehead et al. (1981) for a detailed discussion of these points.
14 See Chapter 7: *Health.*
15 Sen (1987).
16 Sen (1987, p. 27).
17 See Chapter 2: *Statistics.*
18 See Chapter 3: *Agriculture.*

Chapter 9

Practical guidelines

Cecilia Andersen

INTRODUCTION

Women in Development (WID) has been a bona fide issue on the world's agenda since 1975, the United Nations' International Women's Year. Much has been accomplished in the intervening years. Perhaps the most crucial change is that the debate has moved on from considering the 'whys' and 'whethers', to discussing the 'hows'. As the world awakes to what women can contribute as productive and social forces to their countries, the people with the power to make things happen in this world are increasingly confronting the *practical aspects* of improving the quality of aid through a gender-aware approach to all development activities.

For example, for the members of the Development Assistance Committee (DAC) of the Organisation for Economic Co-operation and Development, concern for WID has evolved from 'symbolism to general practice';[1] the issue is now one of its seven priorities. Yet even the DAC members and executive officers would agree that the scope of this issue is enormous and that to date even those donor countries most advanced and experienced with it have only scratched the surface. This is why incorporation of the WID issue in development cooperation must be an ongoing process.

Because different countries and international organizations have approached the WID issue in different ways, we can learn a great deal by studying and analysing their experience. For this chapter we carried out a survey of how the DAC member countries, the European Economic Community and the World Bank have incorporated WID into their aid activities and 'corporate cultures'.[2] Some of the countries have already made substantial progress while others are much less advanced. Nevertheless, their practical experience with the topic allows us to identify useful strategic elements and some general trends.

One of the most impressive general trends is how the Women in Development issue has become one of gender awareness. As experience with projects builds up, policy-makers are gradually realizing that men and women play an overlapping variety of roles which complement each other. A change for one inevitably brings a change for the other. Thus many aid agencies are deciding that a balanced gender-aware approach is the best way to implement development programmes.

This chapter does not seek to delineate the errors and omissions which have occurred with this policy issue. Instead it focuses on the *positive* experiences and achievements of the countries and organizations, in order to give policy-makers, administrators and non-governmental organizations practical guidelines for a step-by-step incorporation of gender awareness into their development-cooperation efforts.

The survey indicates that there are seven basic stages that need to be addressed for a country or organization to deal successfully with the WID issue. These are:

1 Putting gender awareness into the mandate
2 Choosing a policy approach
3 Designing an operational plan
4 Building the administrative structure
5 Dealing with recipient countries
6 Evaluating in gender-specific terms
7 Sharpening the strategies.

By the time the seventh stage is reached, successful administrators recognize they are dealing with a loop. Each stage acts as a motivating or enabling factor that prompts development executives to think about the issue of Women in Development and to incorporate gender awareness into their planning and procedures. There can be simultaneous progress at various stages of the process and work on one aspect can reinforce efforts elsewhere.

PUTTING GENDER AWARENESS INTO THE MANDATE

One of the most essential stages for success is that a clear signal be sent out that the WID issue is a concern at the highest levels of political and administrative decision-making. Ideally parliaments and legislatures should also demonstrate their recognition and acceptance of the issue and build in continued parliamentary monitoring of WID efforts.

The United States Congress provided such a thrust through the 1973 Percy Amendment to the Foreign Assistance Act. It required Bilateral Assistance Programs of the Agency for International Development (AID) 'to give particular attention to those programs, projects, and activities which integrate women into the national economies of foreign countries . . . thus improving their status and assisting the total development effort.'[3] This mandate was broadened in 1977 to recognize the significant economic roles that women play in the economies of the developing countries. Congress sharpened the mandate in 1978 to emphasize that funds for the WID issue be used primarily to support activities that would increase the economic productivity and income-earning capacity of women. In addition, the AID Administration is required to provide Congress with an annual assessment of its WID activities.

Such legislation forces government agencies to work out a strategic programme and to monitor their own progress on the issue. This kind of motivation has obviously worked in the United States. One small example of its success is the worldwide guidance cable which the AID Administrator sent out in 1988 to emphasize the importance attached to WID and to explain the content of related Congressional legislation.

Italy and Denmark are the only other two OECD members to have inscribed the WID issue as priorities through parliamentary action. The 1987 Italian law on development cooperation specifies the need to promote the cultural and social development of women as well as their direct participation in development efforts. In Denmark, women were mentioned from 1975 onwards as specific target groups for development assistance and in 1986 the Danish parliament adopted a resolution calling for a national follow-up of the United Nations Nairobi Conference.

The initiative does not necessarily have to come from parliament. In The Netherlands, for example, growing popular concern for the topic culminated in 1980 in the Minister of Development Cooperation presenting a policy document to the Dutch parliament.[4] This stimulated an ongoing process of parliamentary concern for WID. Now when the annual budget is presented to parliament each year, it offers an opportunity to review the issue and when there are special debates on specific questions the WID angle can be considered.[5]

Political recognition

In many of the other nations governments have boosted the issue's political status by declaring WID a priority. For example, the Australian

Minister for Foreign Affairs took the initiative in 1976 to make a firm commitment to further equality of opportunity for women in developing countries; and in 1983 determined that greater attention would be paid to the effect of Australian aid policy on women in developing countries. A Plan of Action was then formulated to identify objectives and responsibilities over the full range of Bureau programmes.

In 1968 the Swedish Government's proposal to parliament on Development Cooperation declared that high priority would be given to activities for the benefit of women in developing countries; since then budgetary proposals have regularly contained a special reference to the WID issue. In 1980 a Government policy statement reaffirmed its aim that development cooperation reach the poorest segments of the population and stated this included most of the women in the recipient countries. The Government also called for the monitoring of WID aspects in ongoing development programmes and for WID to be systematically incorporated in all future development activities.

In Norway the WID issue was introduced via family planning, which accounted for 10 per cent of its development assistance in 1972 and then via mother and child primary health care. Then in 1975 a Government White Paper for the first time emphasized the need to integrate women more actively in the development process.[6] This led to discussions in the Foreign Committee of the Government and in turn resulted in a first Government proposition on the issue of Women in Development assistance. Women are now a central and explicit target group of development assistance.

In the Federal Republic of Germany the process of recognition of WID was also gradual. It started in 1978 with a Government paper recommending, *inter alia*, the assessment of the direct and indirect effects of development projects on women. This was followed by the publication of a Government paper giving general directives and recommendations on Women in Development. Today, FRG Development Policy Guidelines contain a specific directive concerning the role of Women in Development.

Finland issued a ministerial directive in 1988 encompassing the WID programme of action. Austria took similar action in 1989. The Japanese Government has maintained over the years that it aims to 'ensure the full participation of women in all the processes of development with rights, opportunities and responsibilities equal to men and the assurance of receiving appropriate benefits from development'.[7] In 1990 Japan appointed a senior-level steering committee and a panel of experts to operate as a task force and formulate a WID mandate.

Recognition through administrative channels

Many of the other OECD countries have dealt with the WID issue through administrative guidelines. This procedure has worked very well in Canada, where WID has become one of the country's six major priorities in development efforts.[8] Through years of careful effort, the WID issue was slowly but efficiently made a part of Canadian development cooperation. Canada now has a policy framework which defines the scope of the WID mission assigned to the development agency, as well as the orientation, operational objectives and strategy to be followed.

The United Kingdom, New Zealand, Ireland, Switzerland and France have also chosen to deal with the issue of Women in Development through internal guidelines issued by their ministries or development agencies. Belgium's Ministry has issued a directive recognizing the importance of the issue, but is still in the process of developing a strategy and plan of action.[9]

The international organizations

The World Bank demonstrated its recognition of the WID issue in 1975 by appointing a special adviser whose brief was to increase attention to women in World Bank activities. Twelve years later a small central WID office was established and the World Bank identified Women in Development as one of its Special Operational Emphases. The Bank now has a WID division and special WID units for various regions.

The special supranational structure of the European Economic Community affects the way in which its institutions can approach the WID issue. Concrete EEC actions in the area of WID must be carried out by the Commission, but under a mandate provided by the Council of Ministers. The varying levels of support for WID by Member States affect the way the Development Council deals with the subject and influence the message passed on to administrators. The divergence of views is felt particularly when the Presidency of the Council is passed from state to state every six months, as this has a direct bearing on the priority given to WID during that time period.

Over the years the European Parliament has contributed significantly to making the EEC responsive to the emerging WID issue. As early as 1981, the Parliament expressed solidarity with the women of the Third World and in a formal document declared the necessity for the different sectors of EEC development cooperation to take into account the specific needs of women.[10] This document spurred the first discussion on this issue by the Development Council of Ministers in November 1982. The Council then accepted their first resolution on WID. It expressed concern that cooperation efforts should contribute to the harmonious development of the whole population in the recipient countries and declared that the Community was prepared to take into account the role of Women in Development as well as problems specific to women.

Since that resolution the Council has on various occasions discussed the WID issue. In May 1989 it confirmed that the Community's policy is to take systematic account of women's role in the development projects financed by the EEC. Moreover, the Council has stressed the necessity of completely integrating the WID issue into all sectors of activities and in all agreements concluded by the EEC with its development partners. Going one step further, the Council asked that an action programme be elaborated in order to determine specifically how women's roles are to be taken into account at the different stages of implementing programmes and projects. Then in 1990 the Council requested work programmes to implement the articles of the fourth Lomé Convention that relate to the role of Women in Development.

The Parliament has continued to adopt resolutions aimed at the full integration of women in all development programmes. In 1989, the Parliament opened a budget line for WID. As the political power of this body increases, the weight of its concern for integrating Women in Development programmes can be expected to influence policy more directly.

The EEC's biggest aid programme is with the group of African, Caribbean and Pacific (ACP) countries and is structured in the context of successive Lomé Conventions. Lomé III, concluded in 1985, dealt with the WID issue in the context of cultural and social cooperation. Specific attention to the role of women was linked to operations to enhance the value of human resources.[11]

Lomé IV, which entered into force in 1990, has expanded the provisions on the role of women. It states unequivocably: 'Account shall be taken in the various fields of cooperation, and at all the different stages of operations executed, of the cultural dimension and social implications of such operations and of the need for both men and women to participate and benefit on equal terms.'

It also introduced the new Article 153 on 'Women in Development', which states:

Cooperation shall support the ACP States' efforts aimed at:

(a) enhancing the status of women, improving their living conditions, expanding their economic and social role and promoting their full participation in the production and development process on equal terms with men;

(b) paying particular attention to access by women to land, labour, advanced technology, credit and co-operative organizations and to appropriate technology aimed at alleviating the arduous nature of their tasks;

(c) providing easier access by women to training and education, which shall be regarded as a crucial element to be incorporated from the outset in development programming;

(d) adjusting education systems as necessary to take account in particular of women's responsibilities and opportunities;

(e) paying particular attention to the crucial role women play in family health, nutrition and hygiene, the management of natural resources and environmental protection. Dissemination of information to women and training of women in these areas are fundamental factors to be considered at the programming stage. Appropriate action shall be taken in all operations referred to above to ensure the active participation of women.

The Convention also specifically recognizes women as a vulnerable group within the context of the structural adjustment support to be provided. In addition, the articles describing future collaboration within the agricultural and rural sectors – areas of priority in the Lomé Convention – now include instructions concerning the active participation of the male and female rural population.[12]

The progress made by Lomé IV on the WID issue is notable because it has had to come through a negotiating process with the ACP countries. The increasing emphasis on WID in the Convention reflects an evolution in the ongoing dialogue held in the framework of three joint institutions: the ACP–EEC Council of Ministers, the ACP–EEC Committee of Ambassadors and the ACP–EEC Joint Assembly. The latter institution has contributed specially and extensively to the evolution of the issue through its *ad hoc* working party on WID, in which women from the ACP countries have taken a leading role.

This section has shown the wide variety of ways in which First World aid agencies have given official recognition to the Women in Development issue. But no matter how it has been accomplished, such recognition is still only the initial step. To have real relevance, the WID mandates must also be based on the principle of integration, involving all aspects of the donors' development programmes and all parts of their aid administrations.

CHOOSING A POLICY APPROACH

A realistic starting point for incorporating WID concern into aid activities is recognition of women's triple role and of the distinction between women's practical and strategic gender needs.[13] The term 'triple role' is used because in most low-income households women perform:

1 Reproductive work in the form of childbearing and child-rearing responsibilities
2 Productive work (often as the secondary-income earner), and
3 Community-managing work, which is related to the inadequate state of housing and basic services in most urban and rural areas of the Third World.

As Caroline Moser has so eloquently explained, planners need to be attentive to women's role in the community not only because of the importance of the tasks performed by women, but also because estimates show that one-third of the world's households are now headed by women.[14]

Distinguishing between practical and strategic gender needs

Planning for low-income women should also be based on knowledge of their interests and needs. Here it is necessary to distinguish between 'strategic gender needs' and 'practical gender needs'. Strategic gender needs are those that arise from women's subordination in society and are thus formulated in terms of a more satisfactory organization of society. Practical gender needs are those which arise by virtue of women's position in the existing division of labour. For example, adequate housing or a supply of clean water are practical gender needs of low-income women. Although such needs affect all household members, they are very often identified as 'women's needs' because they relate to women's traditional responsibility for much of the work in the household.

In formulating policy options, aid officials should be aware that there are alternative approaches to the Women in Development issue and that these affect women differently. Five basic models can be recognized: the welfare approach, the equity approach, the anti-poverty approach, the efficiency approach and the empowerment approach.

The five approaches

The *welfare approach* is a social-policy approach for the benefit of a vulnerable group in society. It assumes that women are passive recipients of development, that their role as mothers is their most important role and that child-rearing is the most effective contribution women can make to economic development. Concern for the physical survival of families leads to concern for the provision of goods and services, a focus on the nutritional needs of women and children and, more recently, to family planning. The welfare approach is 'politically safe': it does not meet women's strategic gender needs because their traditional 'natural' position is not questioned. While it deals with some of the practical gender needs of women, the resulting welfare programmes may create dependency.

The *equity approach* recognizes that women are active participants in the development process and in economic progress through their productive and reproductive roles. The original WID approach, especially as formulated by the WID movement in the United States, was an equity one. The basic assumption is that economic strategies often have a negative impact on women and that women should be brought into the development process. Equity approaches meet strategic gender needs and link development with equality. As such they aim at a redistribution of power. In political terms, this makes them less easily acceptable because of their likelihood of interference in a host country's traditions.

This approach also has important methodological difficulties. It requires standards against which progress can be measured – ideally a single, unified indicator of social status or progress paired with baseline data on women's actual situation. Such information can only evolve through the accumulation of disaggregated data, increased research efforts, specific evaluations and collaboration and exchange of information. For these reasons most agencies no longer use the equity approach, but its official endorsement in 1975 by the International Woman's Year Conference still makes it a point of reference.[15]

The *anti-poverty* approach recognizes that for a variety of reasons the majority of women in the Third World fall into the target group which has to be assisted to escape absolute deprivation. The approach emerged when general research on development programmes revealed that accelerated growth strategies by themselves do not solve poverty and unemployment problems, nor lead to a redistribution of income. Institutions such as the World Bank began to recognize that one of the reasons for the failure of the 'trickle down effect' was that development planning had been ignoring women. They posited that one means of alleviating poverty and promoting balanced economic growth was through the increased productivity of women.

Thus the anti-poverty approach focuses mainly on the productive role of women and on the need to provide them with better access to productive resources. Anti-poverty programmes meet practical gender needs if they provide more employment opportunities for women, but the programmes only meet strategic gender needs if they increase women's capacity for self-determination.

The *efficiency approach* concentrates more on the development process and less on women. It emerged when the world economy started to deteriorate seriously after the first oil shock. Women are seen as an underused asset for development; it is assumed that their increased economic activity will – in and of itself – lead to increased equity. In the efficiency approach women's unpaid time is used as self-help components in economic activities, specifically with respect to human-resource development and for the management of community problems. It assumes that women's unpaid labour in areas such as child care, fuel-gathering, food processing, preparation of meals, nursing the sick, etc., is elastic.

Thus the efficiency approach addresses women's practical gender needs, but at the cost of longer working hours and more unpaid work. In some cases it has also been found to undermine their practical needs by using women's unpaid labour to replace reduced resource allocations within structural adjustment policies.[16] For just such reasons this approach fails to meet women's strategic gender needs.

The *empowerment approach* focuses on increasing women's control over the choices in their lives. It seeks to increase their self-reliance and self-confidence so they will become more active players in society. Through increased control over crucial material and non-material resources, women are then expected to take steps to influence the direction of social and economic change.

This approach emerged most forcefully from the writings of Third

World women and the experience of their grass-roots organizations. It does not centre on the domination issue, but rather questions two assumptions of the equity approach: (i) that development necessarily helps all people and (ii) that women want to be integrated into the mainstream of Western-designed development.

The empowerment approach recognizes the triple role of women and views the work of women's organizations and like-minded groups as a key element of change. It champions the use of a 'bottom-up' approach to raise women's consciousness so they can challenge their status in society; it works on practical gender needs to build a support base in order to address strategic gender needs.

Donor agencies often use various approaches simultaneously. In practice, the welfare approach is the oldest model and it has been the one most widely used. The equity model was the original WID approach, but its political sensitivity has prompted a shift to the anti-poverty approach. Yet that approach can also have political connotations because it tends to isolate poor women as a category. If income-generating projects then lead to more income and greater autonomy only for the women, the power balance between men and women may be affected. Another danger here is that if programmes focus too much on the productive role of women and ignore their reproductive role, then the end result may be an extension of women's triple burden.

Problems arise with the efficiency approach because cuts in spending on human resources within structural-adjustment programmes tend to result in a serious deterioration in living conditions for the low-income population. The cuts also have a gender-differentiated, knock-on effect on intra-household resource allocation which is found to be especially detrimental to women and children.

The empowerment approach is important for donor agencies because it has its main roots in the history of Third World women. Yet the approach is difficult for donors to use because of its political connotations and because it emphasizes grass-roots activities and the agencies have to work through official channels.

Those donor agencies that have sought to fashion a coherent policy approach and have met with the greatest success have chosen various elements of the five policy models and then allowed their overall approach to develop pragmatically over time in response to perceived needs and changing perceptions. The challenge which this issue poses for aid policy-makers and planners is to create and design programmes that meet both the practical and the strategic gender needs of women.

DESIGNING AN OPERATIONAL STRUCTURE

Generally countries tend to follow their initial recognition of the WID issue with the drafting of an action plan or a set of guiding principles. The aim of the basic document is to provide priorities and to set strategies for the systematic, step-by-step integration of gender considerations into policies, procedures and projects. Good documents establish benchmarks, timetables and indicators for the monitoring of compliance. As long as the document is regularly reviewed and updated it can continue to be used as the basis for regular reporting on WID to legislative, administrative and political bodies. It can also serve to demonstrate the agency's WID position to aid recipients.

Detailed plans of action now exist for six DAC member countries while 17 have an action programme. Canada, Australia, the United Kingdom and the United States have introduced specific requirements for monitoring the implementation of their action plans.

Various approaches are used simultaneously and over time. Refinements are made as the WID issue becomes incorporated into the activities of an agency. Countries with a longer history of WID activities have gained insight into the complexity of the issue and use this to construct policies that treat the topic in a deeper and more profound way. Some donor countries with a more recent policy for WID, such as Italy, Switzerland and Finland, have benefited from the others' longer experience to construct an approach that was more refined from its inception.

Evidence from the Nordic countries, The Netherlands, Canada and the United States indicates that an in-depth approach is feasible. Their experience also proves that over the years the subject matter gains a momentum of its own, becoming an accepted element of the agency's overall approach. Moreover, the lessons learned in making the WID issue part of daily operations have proved to be useful in introducing other new topics, like the environment, into development assistance.

WID criteria can be incorporated into the general-sector guidelines adopted by an agency. Four countries have chosen to focus on areas of specific importance to women: health, water, sanitation, education and agriculture. It is also possible to take a geographical, country approach. The United States has shown that agencies can deal with the WID issue across-the-board – for all sectors and all countries. Yet another approach is specific WID guidelines, including checklists for sectors where women's role is critical.

The basic documents on WID need to be distributed and made

familiar to staff throughout the agency since they affect ongoing activities. This can be accomplished in various ways. One example is the United States AID Administrator's worldwide guidance cable in July 1988 explaining the meaning of Congressional legislation and how all overseas missions and Washington-based bureaux should respond to it. In The Netherlands, the WID action programme was discussed at different levels of the Administration and then sent to all heads of sections within the Ministry and to every embassy. Norway explained its strategy and plans of action and then followed through with specific instructions. In the EEC a special newsletter on Women in Development is published by and for Commission staff working with development efforts. Some countries have chosen to issue WID manuals, which contain explanations and suggestions for the key sectors of relevance to women. The EEC has developed guidelines on how to incorporate the gender issue in the project cycle. Whatever process is used, the end result should be that the agency's overall WID objectives become part of the regular functioning of the administration.

The survey shows that the countries where a coherent policy approach has been adopted and continuously supported have had the most success in motivating their staff and in creating workable programmes. The DAC now recognizes WID as an issue of fundamental economic importance and one that cross-cuts with relevance for all the activities of development cooperation.

As the United Kingdom strategy paper states: 'The calculation that women account for two-thirds of the world's work hours, receive ten per cent of its income and own one per cent of its assets may be apocryphal but it is probably not too wide of the mark. It indicates the economic benefit to be secured from promotion of women's interests. Women are important agents of development and without their support and cooperation aid activities will be less successful.'[17] One difficulty with such an approach is to ensure that it leads to greater social equity for women but not to an increase in their 'triple' workload.

Australia has sought to safeguard its programmes from falling into this trap by setting the following policy objectives:

- To increase the impact of the development-assistance programme by ensuring the participation of women in planning and implementation
- To ensure the programme's greater effectiveness by taking account of women's needs and preferences
- To assist the development of recipient countries by increasing the productivity of women's activities

- To ensure that women share equally with men in the benefits arising from Australian development assistance.

This kind of policy approach has consequences for the way in which WID is addressed in projects and programmes. If women form part of the target group or are an important element in the workforce used in specific projects, the success of the project might well depend to a great extent on their capacity to function as agents of change. This in turn will depend on their overall workload, their physical health, their motivation and their social status.

When assessing these factors, aid administrators also need to recognize that women's multiple roles are their everyday reality. However, knowledge about these factors is not readily available. Gender-disaggregated baseline data need to be designed so that this information is collected. Through such data, analysts should be able to discern the constraints for various socio-economic, ethnic, age and other groups. Projects can then be designed to overcome the constraints and offer the incentives needed to meet with success. This requires the collection and analysis of disaggregated data in the course of project identification, realization and evaluation. Methodologically this is not an easy task; moreover, it is costly and time-consuming. Yet, in the long run, it is the only way.

The realization that administrators need support for taking this step was the reasoning behind the United States Congressional legislation requiring gender-disaggregated data to be collected in the design, implementation and evaluation stages of all projects. Canada also requires that disaggregated data be used from the initial stages of a project cycle. The same trend can be discerned in other countries where substantial progress is being made with WID.

Social welfare and the economic role of women

Administrators now readily accept that women in developing countries need greater access to nutrition, health services, education, training, employment and financial resources. Yet too often donors still concentrate on the social needs and less often on the economic and empowerment connotations. Recognition of women's needs which is based mainly on the traditional perspective of women as a weaker group can act like a double-edged sword: it can help them to improve their physical and mental ability to take advantage of opportunities, but at the same time it fails to create such opportunities.

For example, Italy has formulated guidelines which have the promotion of the role of women in developing countries as their primary objective. The agency expects specific programmes and projects to consider the role and functions of women and in practice there has been a distinct emphasis on women's needs. Yet the needs of women and children are still placed side-by-side. The fundamental duality that exists with respect to the WID issue can also be illustrated through the example of Sweden, where two strategies are combined. One concentrates on welfare and care, aiming to reach women and children with social and health-oriented measures. The second strategy aims to raise the value and productivity of women's work and to strengthen women's position in the economy of the society in which they live. The Swedish International Development Agency (SIDA) now puts more emphasis on the second strategy.

Denmark's experience can illustrate the evolutionary nature of donor WID policies. In the 1970s women's issues were mainly integrated into projects related to the provision of drinking water, health and education and to a lesser degree into the more directly productive projects (agriculture, fishery, forestry, industry). A dominant trend was the welfare approach, based on a more or less explicit concept of women as victims of a special development process, as marginalized and consequently as receivers, beneficiaries of development assistance. In the 1980s more emphasis was placed on a production-oriented approach. This change in general strategy represents a shift from the basic-needs strategy to one that is more oriented towards economic growth. Women are now seen as socio-economic actors and the central role of women in food-crop production is given special consideration.[18]

The World Bank's perspective on WID illustrates the links that exist between WID policies and the overall line of reasoning which applies to the total aid programme. It is highly influenced by the efficiency policy model and is linked to its work with structural-adjustment programmes. The Bank is convinced that the expansion of women's opportunities can lead to enhanced productivity and earning potential, which will improve women's own living standards and in turn contribute to better economic performance, poverty reduction and family welfare. It is especially committed to this policy for the many poor families who are headed by women. The Bank also expects that over time the build-up of WID activities from all the donor and host countries will help to slow population growth.

The EEC experience with the WID issue is a special case because the Lomé Convention represents an obligatory baseline for action on WID which has been negotiated with the ACP countries. The EEC policy approach lays specific emphasis on the role of women in agriculture and food production. Thus the 1988 progress report and the EEC guidelines for an operational programme emphasize women's rights to access to land, credit, extension services and technology. Women's role is also noted in the preservation of food to reduce post-harvest losses, in livestock development and in fisheries. The guidelines also recognize women as domestic managers and consequently situate women as decision-makers with respect to water supply, sanitation and health care.

Once donor agencies recognize Women in Development as a cross-cutting issue, they must find and/or create the means to have the issue become part of the concrete sectoral and geographic activities of the aid programme. Success at this stage depends not only on the policy approach adopted, but on an agency's willingness to invest in putting the people, funding and staff-time where they are needed.

BUILDING THE ADMINISTRATIVE STRUCTURE

The best way for aid agencies to integrate gender awareness into all their aid activities is to make qualified advisory services on WID available for use by the entire aid administration. But this is not easy to achieve.

Installing a WID focal point

Many agencies have decided to use a WID focal point, i.e. some person or office charged with inserting WID priorities into ongoing programmes and projects. This is not easy either! Often the focal points operate with a very small staff. In some countries the office may consist of a part-time regular staff member, possibly assisted by outside consultants under temporary contracts. Yet the office is expected to be a spearhead in developing overall staff knowledge on WID and to be available to other staff members for advice. The focal-point position also requires them to review and comment upon the work of other personnel in the agency, including those in higher grades.

At the very least, a focal point should be adequately staffed. It is bad management to charge an office with specific tasks and then not allocate sufficient permanent personnel to do the job. Inadequate staffing leads to demotivation because the executives know they cannot perform their task adequately. It may also lead to unjustified negative assessments of their input to the overall work of the administration. A minimum allocation of staff on a permanent basis is vital for successful management of the WID issue. When a country does less, it sends a signal to all its own staff and that of other aid agencies that it is only paying lip service to the WID issue. The United States, Canada and The Netherlands are examples of countries where staffing is extensive.

Whatever the size of the WID staff, the unit should be well-placed strategically. It should be given an advisory, creative and directive role with a direct link to actual policy-making and top administrators. Agencies have different viewpoints on whether direct implementation should be in the hands of only the sectoral and geographic desks. Canada, The Netherlands and the United States are countries where this issue has been broached. Limiting the direct responsibility of the WID unit in day-to-day implementation may have the disadvantage of reducing its power base, since the unit is not involved with line management and related implementation and supervisory activities. If this option is chosen, the support of top management or the existence of an independent reporting mandate – as occurs with the US Agency for International Development – becomes doubly important.

The United States Office for Women in Development provides a wide range of services for AID:

- It provides research and analytic services with a view to establishing a data base and developing analytic capability
- It provides technical assistance in the design, implementation and evaluation of projects and programmes
- It trains the strategic planners and gives mission-specific training
- It trains key project and programme officers, NGO personnel, trainers and contract teams
- It disseminates a steady stream of information on the results of research and analysis, technical assistance and training.

In short, the WID office helps others to fulfil their WID responsibilities. The unit can do all this so effectively because Congress requires the Agency to report its progress on the WID issue.

If a WID focal point is designated, its work can be greatly facilitated by the creation of a WID Inter-Departmental Task Force with members from various sectors of the administration. Such a group need not meet on a frequent basis; the mere fact that it exists and will hold regular meetings makes it possible for the WID issue to be constantly followed up within administrative procedures. This approach is used by a number of aid agencies, including the EEC, and agencies in Australia, Denmark, The Netherlands, Italy, Switzerland, the United States and Canada. Their experience demonstrates that the system works well as long as the activities of the task force receive clear support from top management.

Several countries have decided to integrate the WID issue into their agency by situating it in a department which has a more broadly based mandate.The United Kingdom has positioned its WID focal point in the Aid and Social Policy Group within the Economic and Social Division of the Ministry. Two social-development advisers are specifically designated to provide policy advice on Women in Development and other staff members also have WID responsibilities:

> We regard this as a more effective way of helping women than through creation of a special Women's Unit or allocating special funds for women's projects. With the latter approach there is a danger that the issue will be marginalised. Thus we do not set out to finance projects just for women though many projects may by their nature concern women rather than men (or the reverse). We are also prepared to fund components of projects to ensure that women as well as men benefit.[19]

In addition to their WID focal points at the agency's base, some countries have appointed full- or part-time WID experts in the field.[20] They find this gives them the advantage of direct contact with aid recipients, easier dialogue with official and non-official contacts and on-the-spot observation of gender needs and how the aid policies are affecting them. This then leads to better policies, programmes and projects. Other countries have begun to assign WID-related tasks to some of their overseas missions.[21]

Deciding how to finance WID activities

Once an aid agency has decided upon the number, quality and ranking of the staff it wishes to assign to WID activities, it must then choose how it will finance their activities. There are three methods: (1) integrated funding through the normal budget lines, (2) specific funds for special projects, and (3) a special budget line. The key objective is to ensure the integrated funding. This meets the fundamental aim of incorporating WID into all programmes and projects in the agency. Specific funds can be used as supplements and complements to normal, ongoing budget lines, but should never replace the integrated funding.

For example, the Federal Republic of Germany, Australia, The Netherlands and some other donor countries have earmarked small special funds for women's projects. Their agencies find these funds useful for supporting very small projects proposed by or for women's groups that do not fit into bigger programmes and for reaching target groups that are unattainable through other means. The Netherlands has a special budget line of $2.5 million annually to support WID activities at home and in recipient countries. Such special funds can also be used to finance specific initiatives through multilateral organizations.

One of the best reasons for setting up a special budget line for WID is to enable the focal point to serve as a catalyser. The funds can then be used as seed money, permitting key activities to become embedded in a wider context. The United States Congress has done this by earmarking $5 million annually to assist AID in meeting the requirements on WID, but it has stipulated that the monies are to be used to 'supplement and encourage additional spending for women and expansion of development activities ... not as a substitute for other AID funds that benefit women's development.'[22] The WID office is allowed to use the money to fund fully, co-finance or provide matching funds for contracts and services.

Issuing specific country and sectoral instructions

For the Women in Development issue to be fully institutionalized within an aid agency, it must be integrated in all stages of the work, from programming and planning through evaluation and follow-up. This usually involves the drafting of specific country and sectoral instructions with respect to WID. Norway's aid agency has experience in this area. Its bilateral aid is concentrated on a limited number of countries, so it drew up plans of action for each main partner country. These plans identified particular targets that are measurable and have a specified time frame.

WID experts can help to formulate such benchmark documents. But in the end it is best if executives in all sectors of the administration take final responsibility for formulating and implementing gender-aware action plans for their specific lines of activity.

It is important that donor agencies realize that quantitative and qualitative targets on their own are not sufficient; for the goals to be reached, gender awareness must be incorporated into the regular systems and procedures. WID experts can advise, monitor and evaluate, but the people who are in charge of the project or programme must determine what is feasible and then be held accountable for the results. Some countries are adopting an 'accountability' system. However, for the approach to succeed executives must have adequate knowledge of

WID. An essential step, therefore, is training on the Women in Development issue.

Training in gender awareness

Several countries have already had extensive experience with training activities, so this permits us to identify various important elements. One essential factor is that training be made obligatory for senior levels of administrative personnel.[23] As much as possible the training should be sector or regional specific and adapted to peoples' particular work responsibilities. Moreover, the training should be a continuous process and not a once-in-a-lifetime event.

Canada and the United States have extensive training programmes and are good examples of their benefits. They report that the more they train, the more they find that members of their administrations have become aware of the WID topic, are able to assume a WID responsibility in their areas of activity and serve to engender further awareness in other staff and contractors working with them.

But the content of the training sessions must be well planned to enable such gender awareness to evolve. In the initial stage, training should introduce staff to the WID issue. Subsequent sessions should concentrate on ways and means of incorporating a WID perspective into programmes and projects. At the moment, material is widely available for the first type of training. The second stage requires an input that is much less easy to develop but is now more and more available.

Ideally, such training materials should be adapted to the policy and set of programmes of each agency. The training should also respond to the needs of various audiences. The chosen materials have to offer concrete practical help to individual executives grappling with the application of gender awareness to their day-to-day work. If the agency succeeds in its goals, a third interactive training stage should follow after executives have had experience applying gender-aware principles to their work. They are then more open to discussing with experts and their colleagues the effects of the new, gender-aware way of viewing development and society.

New staff entering aid agencies are in many cases being given gender-aware training as part of their general orientation. This is happening both in that set of countries which have a long tradition with the WID issue[24] and in the set which is benefiting by 'leapfrogging' over the experience of others to fashion a coherent, refined approach from the very start.[25] Such training is important for building a solid base for the future. But responsibility for the present still rests with the older generations and the lion's share of the training budget should be used for them.

The methodology for gender-aware planning, policy implementation and project realization is gradually taking form. Experience with past and ongoing projects will provide insight into what has been happening in the field. But this knowledge must be translated into training materials before it can be used. This will be difficult because conditions in countries and regions differ considerably, because no two projects are alike, because each donor has its own specific systems and procedures and because training must be agency-specific. Notwithstanding these constraints, it is possible for one agency to adapt and use training material developed for another agency. General background material concerning theories, methodologies and case studies can be used to create specific modules. The EEC has made the development of suitable training material an area of budgetary priority. Its WID office funded a group of eight experts to prepare and publish a set of training materials which could be made available to aid administrators, researchers, NGOs and students throughout the First and Third Worlds.[26]

The World Bank over the past 15 years has concentrated on the compilation of operational examples to demonstrate to Bank officials that WID efforts are practical and make economic sense. The Bank has also worked out a rationale and a conceptual framework for WID and provided training activities for its officials. The results of these activities are becoming apparent.

A recent review of the treatment of WID issues in the preparation and design of World Bank lending operations showed that one in five operations approved in fiscal year 1989 included project-specific WID recommendations compared to one in ten operations in fiscal year 1988. The Bank's activities to assist women are mainly concentrated in priority sectors such as education, population-control, health nutrition and agriculture. The review showed that ongoing projects had been modified during implementation in order to permit greater attention to be paid to women.

Thus training on WID issues has proven its worth: it produces results and spin-offs. But the quality and quantity of training materials available affect the level of training that can be offered. More emphasis needs to be put on the creation of good materials. This is an area in which aid administrations can certainly collaborate.

Outside contractors

Gaining a WID perspective is sometimes more difficult for outside contractors than for permanent staff members who are exposed daily to their agencies' corporate cultures. This can have a crucial bearing on projects that are handled by these contractors. Experience in a limited number of countries indicates that terms of reference should mention Women in Development. Australia, for example, has ensured that such references are now included in its terms of reference for consultant exercises and it has issued consultants with the appropriate sections of the Country Programme Operations Guide.

Some agencies require firms to give proof of expertise in WID. In such cases contractors are obliged to have a staff member who specializes in the topic or to resort to other solutions. In countries such as the United States – where the WID mandate is applicable for all segments of development-cooperation activities and where regular reporting is required – it is easier to work out a strategy for outside contractors than in countries where the WID mandate has less solid underpinning.

There are clear indications that quite a number of countries have begun to extend WID to private enterprise contractors. The FRG has pointed out to consulting firms that greater knowledge and know-how on WID will be expected of them. Vrouwenberaad, the leading Dutch pressure group on WID, commissioned a survey on private-sector interest in the issue and gave it the telling title *Women and Development Means Business.*[27] Finland has already initiated training for consultant firms.

Extending WID to multilateral activities

Donor countries must also take steps to ensure that their WID objectives become an integrated part of their activities in multilateral development organizations.[28] A useful step in this process is the drawing-up of a basic document – be it an action plan or a policy paper – which can be distributed and consulted. The aim here is to generate a consensus on basic objectives and strategies. It should then be possible to insist that WID becomes an element in formal board meetings as well as informal consultations with multilateral agencies.

Although by now the major multilateral development organizations have all expressed concern for WID, the way in which this commitment is translated into practice varies substantially. There is a series of positive steps a donor country can take to influence this development. Several countries have instructed their delegates to support statements and resolutions in favour of Women in Development, but more can be done.

Canada, for example, monitors on a systematic basis the WID performance of the multilateral agencies of which it is a member. Donor countries can also support special activities and projects that are of interest to WID. Some donor countries support the WID efforts of a specific international organization by providing an additional member of personnel. Another positive strategy is to share information on specific topics of mutual concern and to use this information to construct a common lobbying platform. For example, in negotiations leading to the Fourth Lomé Convention, some EEC countries decided to lobby together for greater recognition for the WID issue.

Expanding the work of the WID Expert Group of the OECD's Development Assistance Committee would be another positive step in the multilateral domain. The DAC already provides a channel for the systematic exchange of information on how WID is being implemented by donor countries. In 1983 the group prepared the DAC Guiding Principles to Aid Agencies for Supporting the Role of Women in Development and subsequently revised these in 1989. It has also prepared three monitoring reports on their application. The Group has also been instrumental in arranging for WID to form a regular element of the reporting procedure of donor countries to the DAC. But the donor countries could decide to use the OECD even more extensively as a structured-information exchange. At the moment donors do not know enough about each others' work and about the resources that are available in the various agencies. This leads to duplication and inhibits a mutual learning process. Moreover, it is expensive and not efficient.

To solve this the OECD Development Center could be asked to create a WID research-referral unit. At the moment, each donor country is building up its own database on the issue. This will incorporate the work done by the agency, thus reflecting the degree of investments made by the agency in the WID issue, as well as material from sources outside the agency, such as research institutes and NGOs in the country and abroad. In some countries this is leading to a substantial database, but others lack the personnel and financial capacity to generate such an effort. It would be financially interesting for all the donor countries to fund a computerized clearing-house of information that would give them access to each other's WID country or sector background papers, training manuals, evaluations, etc. Such an initiative could also be very valuable for universities, research institutes and NGOs the world over. Much of the material on WID now circulates through channels that are not sufficiently funded or structured to make them accessible to the people who need them.

Integrating NGOs in WID activities

Donor-country development NGOs have become increasingly interested in the issue of gender awareness. This is reflected in the growing number of projects presented for co-financing in which WID is incorporated. But one consequence of this growing concern is that NGO administrations are experiencing managerial problems with the WID issue which are very similar to those faced by donor agencies. By sharing information and 'lessons learned', agencies can help NGOs construct their own strategic and practical approaches to integrating gender awareness fully into their activities. The NGOs also need to be kept well informed of the official positions of donor countries on the WID issue and of the progress being made in aid agencies.

The Netherlands achieves this by having regular consultations with NGO officials. Australia, the EEC and the United Kingdom now incorporate WID in their approval procedures for NGO projects. The UK allows NGO officials to take part in training sessions organized by its agency. Other countries, including small donors like Finland, organize training specifically for NGOs. Some countries and organizations also sponsor special meetings for NGOs. For instance, the 25th EEC NGO General Assembly had as its major theme 'Women and the Lomé Convention'. The EEC is also helping to establish a permanent WID network among the NGOs.

These actions demonstrate the agencies' recognition that outside pressure groups can be a stimulating force for continuous progress in the official WID approach. The Dutch Ministry of Development Co-operation funds Vrouwenberaad, a network group which exercises a lobbying and a watchdog function. Italy sponsors a similar initiative. In Sweden, SIDA has appointed an advisory women's council of 11 members from political and other national women's organizations. In Finland, the Advisory Board on Economic Relations between Finland and the Developing Countries (TALKE) deals with special WID topics presented to it by a wide coalition of political and women's organizations.

In addition to supplying NGO groups with information on WID issues, aid agencies have to develop the knowledge of the general public on this subject. Publication of the basic mandate and special progress reports are useful starting points, but they should be supplemented by other activities where possible. The reasoning behind this suggestion is that public support for the WID issue and pressure for its continued development are critical elements in its continued success, but the public needs information before it can exercise pressure. Ireland is one of the countries that recognizes this. Although it has a modest development-aid programme, Ireland deals with WID issues in its annual report and in all its departmental publications.

DEALING WITH RECIPIENT COUNTRIES

Most donor countries have communicated to recipient countries their ongoing interest in the WID issue. Initially this is mainly done through a distribution of relevant documents. The next step is to insist that WID becomes a regular element of discussion in the annual bilateral consultations. Most countries have institutionalized this process through the inclusion of the WID issue in annual programme reviews.

But donors can go further. The United States, The Netherlands, Sweden and the World Bank have WID personnel in the field. Sweden has linked this approach to the development of action programmes that facilitate the integration of WID into its bilateral-aid programmes. SIDA began in 1986 with the establishment of a regional programme for Africa. For the next two to three years the bulk of WID office activities were moved out of the Stockholm Headquarters and centred in Nairobi. Now the regional office will be closed down and WID officers will be employed in the local SIDA offices.

The argument that recipient countries are reluctant to respond to donor WID interest and do not propose the right types of project initiatives has been proved invalid. Sweden, the United States, The Netherlands, the EEC (within the context of the ACP Convention) and the World Bank all report a constantly expanding mutual interest in WID projects. Sweden summarizes its experience as follows:

> Indeed, it would appear that when a professionally trained person is designated for the specific task of seeking ways and means for integrating women in development programmes, policy-makers in the recipient countries may be more than willing to accept this principle as an immutable factor in the bilateral negotiations and to back up that acceptance by bringing in their own country's women as participants in the negotiations. In both Tanzania and Ethiopia, the

> authorities have been far more open to initiatives along these lines than was expected. Perhaps the problem in the past was not so much one of lack of will as the lack of tangibility.[29]

The implication is that donor agencies can in future shape their WID strategies and expect governments of the developing countries to contribute methods of how best to realize women's economic potential within their respective socio-economic contexts. Recipient countries should also be expected to propose aid initiatives that take account of the practical and strategic gender needs of women. For this, the recipients' officials should be asked and expected to consult with women's groups and other NGOs in their countries when setting priorities and designing programmes.

Donor agencies should develop a close consultation process with women in developing countries. These women are an important sounding-board; listening to their opinions ensures that an agency's WID strategies address both strategic and practical gender needs. The sharpened perceptions of donor-agency staff should in turn influence their discussions and negotiations with recipient countries. Some donor countries make a special effort to help women in developing countries to increase their capacity to function as agents of change. The Unites States has funded a project in which curricula and teaching materials have been developed and distributed to women's organizations in places such as Lesotho, Malawi, Thailand and the Philippines.

Donor countries can also provide WID training for members of the administration of recipient countries. Such initiatives have become a regular part of the WID activities of the United States and Canada. This training has the double advantage of improving the chances of getting WID issues introduced on the agendas in bilateral negotiations and of helping recipient countries to understand donor-country WID objectives.

Recipient countries should also be encouraged to increase the proportion of women participating in training in donor countries. Quantitative targets are not sufficient here. Donors should insist that women be given access to all areas of training and not be limited to those traditionally associated with females.

Funding WID research with recipient countries is an aid effort that provides triple benefits. It strengthens the WID research capacities of recipient countries, facilitates the gathering of necessary data and can cost less. The United States, Canada and The Netherlands are three countries which already sponsor joint research on WID. In fact, 18 per cent of the research budget of the Canadian aid agency is allocated to such studies.

EVALUATING IN GENDER-SPECIFIC TERMS

Evaluation of programmes and projects in gender-specific terms can play a key role in motivating an aid-agency's staff to take a gender-aware approach and thus achieve its policy goals. Since evaluations are costly and time-consuming, this is another area for structured collaboration among donor administrations. Such pooling of resources enables every country to 'leapfrog' to an extent, by benefiting from the experiences of others.

A database is beginning to build up, for aid agencies are increasingly evaluating their projects and activities in terms of WID. The EEC, for instance, has financed a thematic WID evaluation of ten mainstream development projects financed by the European Development Fund. The Netherlands, Canada, the United Kingdom, Denmark, Sweden and the United States have also been active in the area of evaluation. For the past five years the World Bank has systematically assessed the impact of projects on women in agricultural production. This is done within the context of the Bank's Impact Evaluations, through which past projects are examined five to ten years after their completion.

Some of the evaluations are desk research, but extensive field evaluation is also carried out. The composition of the teams for these reports is important. Unfortunately, at this point in time, although most WID action plans require evaluations to investigate the impact of projects on women, the team members assigned to the job rarely have WID competence and experience. The work requires skilled personnel, funding and time – and all three elements are in short supply. It would therefore be very useful if the results of evaluation activities were disseminated on a much wider scale than they are today. An equally valid reason for collaboration is that the results of the research need to be used systematically as a basis for decisions on future activities. The World Bank disseminates the results of its WID research through its Working Paper Series.

At present most WID evaluation activities by necessity very much reflect past lack of concern for WID. Projects in which women are involved but which were planned without taking women into account can provide information on the consequences for the success of the project. Such evaluation reports can also be used to sensitize development workers to the need for a WID perspective. But to build on our expertise with this issue we need to evaluate and document projects which have taken women's role into account from the initial policy and planning stages. The more the economic importance of women and their capacity to act as agents of change is documented, the more credible the WID issue will become to development executives.

A crucial first step for obtaining a continuous stream of relevant information is to have the baseline data for projects sex-disaggregated. Many agencies are already tending towards this approach. If it were included in the evaluation of all development projects, we should soon have a valuable data resource showing the difference in projects implemented with and without gender-differentiated data and a WID perspective.

We shall need both quantitative and qualitative analysis to help us discern what can be obtained through such an approach. This means that donor countries should be supporting fundamental as well as practical research. Agencies should not only contract-out research to independent consultants to meet specific needs, but also support institutionalized research on WID in universities and research centres.

Aid administrations should help to motivate universities and research institutes to develop their research skills in this area, to perform the theoretical and applied research that is needed so badly and to include training on gender awareness in the curricula. Agencies need agronomists with specific knowledge on WID, demographers with special WID knowledge, economists specialized in WID, etc. Fifteen governments are already trying to spur academia to meet these needs by commissioning research.[30] In the United Kingdom, the United States, Canada, The Netherlands and some other countries, Women in Development is slowly but steadily becoming a recognized subject for established research and teaching institutions.

SHARPENING THE STRATEGIES

As data and evaluation studies begin to demonstrate the benefits of integrating women into development efforts, and as general support for the WID perspective increases, aid agencies seek to improve upon their previous positions. Often the original mandate is reviewed and strengthened, as was done by Australia, Austria, Italy, The Netherlands, the UK and the USA.

Through training and experience planners obtain a better appreciation of what a gender-aware approach implies for their activities, so they ask for increasingly specific and refined operational plans. The strategic role and responsibilities of the WID focal point or experts increase. This should lead to increased investments in personnel. The WID experts can then improve their support functions, providing further stimulation for continuous improvement refinement and the fine-tuning of gender-awareness throughout the aid structure.

This fine-tuning leads in turn to the introduction of a gender perspective to aid activities which traditionally have not been considered relevant to women. For example, the importance to women of education, health, sanitation and environmental projects is now considered evident to all in both donor and recipient countries. Most people now also accept that women play an important role in subsistence agriculture and should benefit from agricultural projects. It is less obvious to the same people that large industrial projects may also affect women.

Fortunately, progress is being made. The United Kingdom's 1988 Strategy Paper contains the following passage:

> There will be less scope for design to benefit women in many infrastructure projects. But such projects should not be overlooked in considering WID interests. For example, households are main beneficiaries in rural water and sewerage schemes and the particular needs of women should be taken into account in their design. A similar point applies to the design of rural roads. Tariffs in electricity schemes sometimes discriminate against households in favour of industry and commerce and therefore go against women's interests. The possibility of a WID dimension in infrastructure schemes should always be explored even if it is then dismissed.[31]

Gradually but inevitably a successful WID strategy leads to the acceptance that women are relevant to all projects in development cooperation. They represent half the population of every country. But they are relevant to different projects in different ways. Proper training, evaluation and research will gradually lay the foundations that will persuade and motivate aid administrators to broaden their WID strategy to include all their agencies' activities.

CONCLUSIONS

The issue of gender and development forms part of the broader issue of the quality of aid. The emergence of the WID issue and the growing recognition that gender-aware approaches are socially as well as economically justified are leading to greater concentration on the quality aspects of aid.

The effective integration of the Women in Development issue in official development cooperation efforts is a gradual process. Experience has shown that success depends on the extent to which the whole administration proclaims formal recognition and responsibility for the issue. This conscious acceptance must then be followed up through the whole chain of project-cycle decision-making. Executives can only achieve this if they are given sufficient exposure to the topic through training.

Once a commitment is established, agency personnel at all levels need access to facilities that offer competence in the WID issue, combined with competence in the various social sciences. WID staff have to be given sufficient funds, personnel and time to influence the entire agency's approach to the development of human resources.

The integration of gender awareness requires an intensified collection of gender-specific data. It is only with such data that gender-aware planning, implementation and evaluations are possible. The gender-aware approach should not be restricted to specific projects, but extended as much as possible to non-project aid. The WID activities of NGOs, research institutes and universities also provide valuable input and should be given continuous support. Only by increasing and extending our information base can the effects of a gender perspective on donor projects and objectives become better documented and understood.

WID has been recognized as a cross-cutting issue. It is slowly being positioned as such. The process can and should go faster. Top-level political and administrative decisions can make this happen. Experience with gender-aware integrative strategies provides valuable information for other cross-cutting issues, such as the environment, population growth and participatory development. In a similar way, WID methodology can be developed with reference to experience with other cross-cutting issues. For example, environmental assessments needed to meet reporting requirements on specific issues have shown how vital it is to have detailed benchmark data.

Recipient countries should be increasingly encouraged to take the lead in determining the best WID approaches for their populations. This requires both recipient and donor countries to study the WID perspectives of the programmes and policies presented for aid cooperation. The experience of the Lomé IV Convention provides an example of such a dialogue.

Lasting and substantial progress on WID will depend to a great extent on ongoing training. The cross-cutting dimension of the issue makes it essential that a consciousness on WID becomes a natural and regular part of operations in both donor and recipient agencies concerned with development cooperation. This is only possible if the personnel are informed and trained.

The gender and development issue provides a unique opportunity for progress on the quality of life. Theodore N. Schultz, on accepting the 1979 Nobel Prize in Economics, stated: 'We have learned that poor people are no less concerned in improving their lot and that of their children than those of us who have incomparably greater advantages. Nor are they any less competent in obtaining the maximum benefit from their limited resources.'[32] The majority of the women in the Third World belong to this category of poor people. They have the right to expect that development aid will also improve their competence to improve their lot.

We have spoken here of 'the WID issue', using the acronym and jargon terminology of aid administrators. This is natural and normal, for in our work we need mental and verbal shorthand to condense such complex issues to packages we can deal with. The jargon only stands in our way when, as policy-makers and administrators, we forget the people affected by the problem.

This issue of gender and development is indeed vast. It marks the lives of every human being and foetus in the developing world. Many years will pass before we see widespread progress that can be measured on a quantifiable scale. We shall have to struggle not to become discouraged, but to remember that each step – however small – taken by policy-makers, planners, project directors, consultants and NGOs will move us forward along the path. In striving to take those small steps we should also remember that by seeking to address this issue, we in the First World become better human beings ourselves.

NOTES

1 OECD Development Cooperation Report 1987, Paris (1988, p. 26).

2 The members of the OECD Expert Group Women in Development and the OECD secretariat have provided valuable support for this chapter. In 1990 they provided information on what is taking place in the various OECD countries. The Group prepared the third Monitoring Report on the Implementation of the DAC Revised Guiding Principles on Women in Development (1989). This was approved by the Development Assistance Committee in 1991. It was made available to the general public in 1992 as OCDE/GD (g2) 17.

Although other international organizations, such as the United Nations Food and Agricultural Organisation, the World Health Organization and the UN Industrial Development Organisation, have also included a WID perspective in their activities, their initiatives are not covered here. The EEC and World Bank were chosen as subjects because the EEC participates in the meetings of the OECD Women in Development Expert Group, and the World Bank attends as an observer.

3 Section 113 of the Foreign Assistance Act.

4 Andersen and Baud (1987).

5 The Netherlands now recognizes the WID issue as one of the four strategic topics with their own budgetary implications and with relevance for all other aid activities. The other three areas are the environment, Research and Development and urban poverty.

6 Government White Paper No. 96.

7 OECD (1988) Second Monitoring Report on the Implementation of the DAC Guiding Principles to Aid Agencies for Supporting the Role of Women in Development. (This report covers papers for 1985–7.)

8 Canada's other five priority issues are poverty alleviation, structural adjustment, environmental protection, food security and energy.

9 The United Kingdom has always taken an integrated approach to the issue of women and development. But in 1988 it issued the document, 'A strategy for implementation of ODA's policy on Women in Development', containing an explicit mandate for supporting WID programmes. Prior to this, executives worked on the basis of an Office Guidance Note dating from 1978 and an Office Procedure Checklist first issued in 1984.

Ireland's Development Cooperation Division within the Department of Foreign Affairs began to develop a specific concern for WID in 1986. The country's small development-assistance programme does not use explicit outside mandates, so its guidelines on WID take the form of an internal document.

Finland in 1988 adopted guidelines for including Women in Development cooperation and it now plans to formulate a more concrete action programme.

New Zealand's Guiding Principles for bilateral official development assistance mention the WID issue and its ministry uses country strategy papers for dealing with the participation of women.

Switzerland uses internal guidelines and indicators and is moving towards the adoption of a plan of action.

France works with internal guidelines within the Ministry of Development Co-operation.

10 Women and Development in Community Practice, The Impact of Community Actions on Women, COM VIII/149/82.

11 Article 23 §1 stated that: 'Co-operation shall support the ACP States' efforts aimed at enhancing the work of women, improving their living conditions, expanding their role and promoting their status in the production and development process.' §2 stated that: 'Particular attention shall be given to access by women to all aspects of training, to more advanced technology, to credit and to co-operative organizations, and to appropriate technology aimed at alleviating the arduous nature of their tasks.'

12 See Articles 13, 42, 46 and 48.
13 Moser (1989b).
14 See Chapter 5: *Housing*; and Moser (1989b).
15 Moser (1989b).
16 Sara-Lafosse (1984); Cornia et al. (1987, 1988).
17 United Kingdom Strategy Paper (1989, p. 1).
18 Andersen and Baud (1987).
19 United Kingdom Strategy Paper (1989, p. 5).
20 Canada, Denmark, The Netherlands, Sweden and the United States.
21 Finland, the Federal Republic of Germany, Norway, Switzerland and the United Kingdom.
22 AID Program for Women in Development (1990, p. 5).
23 Australia, Canada, Denmark, The Netherlands, Norway, Sweden, the United Kingdom and the USA have run training programmes scince 1984. Australia, The Netherlands and Norway have now made WID training compulsory for some categories of staff, especially senior management.
24 Australia, Canada, the United Kingdom and the USA are making significant progress in gender-related training.
25 Finland, the Federal Republic of Germany, New Zealand, Switzerland and Italy have introduced training programmes to raise staff awareness of gender issues in conjunction with significant changes in their WID strategies.
26 The Institute of Development Studies, University of Sussex, Brighton, published in 1990 a series of seven abstracts which are each modules for training in gender and Third World development. They can be obtained from the University Publications Office, Brighton (ISBN 090-371555-4).
27 Vrouwenberaad Nederlandse Ontwikkelingsinstanties (1988).
28 Fifteen countries have already done so.
29 Andersen and Baud (1987, p. 211).
30 Australia, Canada, Denmark, Finland, France, the Federal Republic of Germany, Ireland, Italy, The Netherlands, Norway, Sweden, Switzerland, the United Kingdom, the USA and the European Community.
31 United Kingdom Strategy Paper (1989, p. 5).
32 Schultz (1980).

References and further reading

1 GENDER

References

Bleie, T. and Lund, R. (eds) (1985) *Gender Relations: The Missing Link in the Development Puzzle, A selected and annotated bibliographic guide to theoretical efforts and South Asian experiences*, Bergen, DERAP Publications, No. 184.

Boserup, E. (1970) *Women's Role in Economic Development*, New York, St Martin's Press.

Whitehead, A. (1979) 'Some Preliminary Notes on the Subordination of Women', in Young, K. (ed.) et al. *IDS Bulletin*, Vol. 10, No. 3, pp. 10–13.

Whyte, S. R., Østergaard, L., Jespersen, C. B. and Steen, A.-B. (1987) *Women in DANIDA-supported Development Projects: An Evaluation*, Copenhagen, DANIDA.

World Bank (1989) *Sub-Saharan Africa: From Crisis to Sustainable Growth*, Washington, DC, World Bank.

2 STATISTICS

References

Anker, R. (1980) 'Female Labour Force Participation in Developing Countries: A Critique of Current Definitions and Data Collection Methods', *International Labour Review*, 122(6), Geneva, ILO.

Baster, N. (1981) 'The Measurement of Women's Participation in Development: The Use of Census Data', Discussion Paper No. 159, Institute of Development Studies, Brighton, England.

Beneria, L. (1982) 'Accounting for Women's Work' in Beneria, L. (ed.) *Women and Development*, New York, Praeger.

Berry, A. and Sabot, R. H. (1981) 'Labour Market Performance in Developing Countries: A Survey' in Jolly, R. and Streeten, P. (eds) *Recent Issues in World Development*, Oxford, Pergamon.

Boserup, E. (1970) *Women's Role in Economic Development*, London, Allen & Unwin.

Deere, C. D. (1982) 'The Division of Labour by Sex in Agriculture: A Peruvian Case Study', *Economic Development and Cultural Change*, 30(4).

Dixon-Mueller, R. (1985) 'Women's Work in Third World Agriculture', *Women, Work and Development*, No. 9, Geneva, ILO.

Goldschmidt-Clermont, L. (1982) 'Unpaid Work in the Household: A Review of Economic Valuation Methods', *Women, Work and Development*, No. 1, Geneva, ILO.

International Centre for Research on Women (ICRW) (1980) 'The Productivity of Women in Developing Countries: Measurement Issues and Recommendations', Washington DC Office of Women in Development, USAID.

King, E. and Evenson, R. (1983) 'Time Allocation and Home Production in Philippine Rural Households' in Buvinic, M. et al. (eds) *Women and Poverty in the Third World*, Baltimore, Johns Hopkins University Press.

Massiah, J. (1981) 'The Case of the Commonwealth Caribbean' in UNESCO *Socio-Economic Studies*, No. 3, *Women and Development: Indicators of their Changing Role*, UNESCO.

Mueller, E. (1982) 'The Allocation of Women's Time and its Relationship to Fertility' in Anker, R., Buvinic, M. and Youssef, N. (eds) *Women's Roles and Populations Trends in the Third World*, Geneva, ILO.

Safilios-Rothschild, C. (1982) 'The Persistence of Women's Invisibility in Agriculture: Theoretical and Policy Lessons from Lesotho and Sierra Leone', Centre for Population Studies Working Paper No. 88, New York, Population Council.

Sen, G. and Sen, C. (1985) 'Women's Domestic Work and Economic Activity: Results from the National Sample Survey', *Economic and Political Weekly*, Vol. XX, April.

Standing, H. (1985) 'Women's Employment and the Household: Some Findings from Calcutta', *Economic and Political Weekly*, Vol. XX.

Thorner, A. and Ranadive, J. (1985) 'Household as a First Stage in a Study of Urban Working Class Households', *Economic and Political Weekly*, Vol. XX.

United Nations Educational, Scientific and Cultural Organization (UNESCO) (1981) *Women and Development: Indicators of their Changing Role*, *Socio-Economic Studies*, No. 3.

Further reading

Agular, N. (1980) 'Women in Labour Force in Latin America – General Report', Seminar on Women in the Labour Force in Latin America, Instituto Universitario de Pesquisas do Rio de Janeiro, Rio de Janeiro, 1978.

Blacker, J. G. C. (1978) 'A Critique of International Definitions of Economic Activity', *Population Bulletin of Economic Commission for West Asia*, No. 14.

Blacker, J. G. C. (1980) 'Some Further Thoughts on the Definition of Activity and Employment Status', *Population Bulletin of Economic Commission for*

West Asia, No. 19.

Buvinic, M., Lycette, M. A. and McGreevey, W. P. (eds) (1983) *Women and Poverty in the Third World*, Baltimore, Johns Hopkins University Press.

Chen, M., Huq, E. and D'Souza, S. (1981) 'Sex Bias in the Family: Allocation of Food and Health Care in Rural Bangladesh', *Population and Development Review*, 1(2).

Datta, G. and Meerman, J. (1980) 'Household Income or Household Income Per Capita in Welfare Comparisons', World Bank Staff Working Paper No. 378, Washington DC.

Deere, C. D. and Leon de Leal, M. (1979) 'Measuring Rural Women's Work and Class Position' in Zeldenstein, S. (ed.) *Learning about Rural Women*, Special Studies in Family Planning, 10(11–12), New York, Population Council.

Dixon, R. (1982) 'Women in Agriculture: Counting the Labour Force in Developing Countries', *Population and Development Review*, 8(3).

Dixon, R. (1983) 'Land, Labour and the Sex Composition of the Agricultural Labour Force: An International Comparison', *Development and Change*, 14(3), Sage Publications.

Evans, A. and Young, K. (1988) 'Gender Issues in Household Labour Allocation: The Case of Northern Province, Zambia', ODA/ESCOR Research Report, Institute of Development Studies, Mimeo, Brighton, England.

Evenson, R., Popkin, B. and Quizon, E. K. (1980) 'Nutrition, Work and Demographic Behaviour in Rural Philippine Households' in Binswanger, H. (ed.) *Rural Household Studies in Asia*, Singapore University Press.

Food and Agriculture Organization (FAO) (1986) 'Programme for the 1990 World Census of Agriculture', *FAO Statistical Development Series*, No. 2, Rome, United Nations.

International Labour Office (ILO) (1976) 'International Recommendations on Labour Statistics', Geneva.

McSweeny, B. (1979) 'Collection and Analysis of Data on Rural Women's Time Use' in Zeldenstein, S. (ed.) *Learning about Rural Women*, Special Studies in Family Planning, 10(11–12), New York, Population Council.

Mitra, A., Pathak, L. P. and Mukerji, S. (1980) *The Status of Women: Shifts in Occupational Participation 1961–1971*, Indian Council of Social Science Research, Jawaharial Nehru University Study, New Delhi, Abhinav Publications.

Mueller, E. (1984) 'The Value and Allocation of Time in Rural Botswana', *American Journal of Agricultural Economics*, 58.

Pittin, R. (1982) 'Measuring Women's Work: The Case of the Nigerian Census', ILO World Employment Programme, Population and Labour Policies Programme, Working Paper No. 125.

Recchini de Lattes, Z. and Walnerman, C. (1979) 'Data from Censuses and Household Surveys for the Analysis of Female Labour in Latin America and the Caribbean: Appraisal of Deficiencies and Recommendations for Dealing with Them', United Nations Economics and Social Council, CEPAL Economic Commission for Latin America.

Safilios-Rothschild, C. (1986) 'Socio-Economic Indicators of Women's Status in Developing Countries, 1970–80', New York, Population Council.

Sen, A. and Sengrupta, S. (1983) 'Malnutrition of Rural Indian Children and the Sex Bias', *Economic and Political Weekly*, XVIII.

United Nations Department of Economic and Social Affairs (UNDESA) (1974) 'Towards a System of Social and Demographic Statistics', *Studies in Methods*, Series F, No. 18, New York.

United Nations Department of International Economic and Social Affairs, Statistical Office, International Research and Training Institute for the Advancement of Women (INSTRAW) (1984a) 'Compiling Social Indicators on the Situation of Women', *Studies in Methods*, Series F, No. 32, New York.

United Nations Department of International Economic and Social Affairs, Statistical Office, International Research and Training Institute for the Advancement of Women (INSTRAW) (1984b) 'Improving Concepts and Methods for Statistics and Indicators on the Situation of Women', *Studies in Methods*, Series F, No. 33, New York.

United Nations Educational, Scientific and Cultural Organization (UNESCO) (1981) 'Women and Development: Indicators of their Changing Role', *Socio-Economic Studies*, No. 3.

United Nations Research Institute for Social Development (UNRISD) (1980) 'Monitoring Changes in the Condition of Women – A Critical Review of Possible Approaches' by Palmer, J. and von Buchwald, U., Geneva.

United Nations Secretariat (1980) 'Sex-Based Stereotypes, Sex Biases and National Data Systems' (ST/ESA/STAT/99).

World Health Organization, Division of Family Health (WHO) (1980) 'Health and the Status of Women', FHE/80.1.

Youssef, N. H. (1980) 'Sex-Related Biases in Census Counts: The Question of Women's Exclusion from Employment Statistics' in *Priorities in the Design of Development Programs: Women's Issues*, USAID, Bureau of Development Support and ICRW, Washington.

Zurayk, H. (1984) 'Women's Economic Participation' in Zurayk, H. and Shorter, F. (eds) *Population Factors in Development Planning*, New York, Population Council.

3 AGRICULTURE

References

Beneria, L. (1981) 'Conceptualising the Labour Force: The Under Estimation of Women's Economic Activities', *Journal of Development Studies*, 17(3).

Beneria, L. (1982) 'Accounting for Women's Work', in Beneria, L. (ed.) *Women and Rural Development: The Sexual Division of Labour in Rural Soweto*, New York, Praeger.

Boserup, E. (1970) *Women's Role in Economic Development*, London, Allen & Unwin.

Boulding, E. (1983) 'Measures of Women's Work in the Third World: Problems and Suggestions', in Buvinic, M. (ed.) *Women and Poverty in the Third World*, Baltimore, Johns Hopkins.

Bryson, J. C. (1980) 'The Development Implications of Female Involvement in

Agriculture: The Case of Cameroon', MA Dissertation, University of Manchester.

Buvinic, M. (1984) 'Projects for Women in the Third World: Explaining their Misbehaviour', Mimeo, Office of Women in Development, USAID.

Caplan, A. P. (1984) 'Cognitive Descent, Islamic Law and Women's Property on the East African Coast', in Hirschon, R. (ed.) *Women and Property: Women as Property*, London, Croom Helm.

Clark, C. M. (1981) 'Land and Food, Women and Power in Nineteenth Century Kikuyu', *Africa*, 50.

Cloud, K. (1976) 'Report of Fact Finding Trip to Niger, Mali, Senegal and Upper Volta', Mimeo, Washington DC Office of the Sahel Francophone West Africa Affairs, USAID.

Conti, A. (1979) 'Capitalist Organization of Production through Non Capitalist Relations: Women's Role in a Pilot Settlement Scheme in Upper Volta', *Review of African Political Economy*, 15–16.

Dauber, R. and Cain, M. L. (eds) (1981) *Women and Technological Change in Developing Countries*, Boulder, Westview.

Dey, J. (1981) 'Gambian Women: Unequal Partners in Rice Development Projects', in Nelson, N. (ed.) *African Woman in the Development Process*, London, Frank Cass.

Dey, J. (1982) 'Development Planning in the Gambia: The Gap between Planners' and Farmers' Perceptions, Expectations and Objectives', *World Development*, 10.

Dixon, R. B. (1980) 'Assessing the Impact of Development Projects on Women', AID Program Evaluation Discussion Paper 8, Office of Women in Development, USAID.

Dixon, R. B. (1982) 'Women in Agriculture: Counting the Labour Force in Developing Countries', *Population and Development Review*, 8.

Dixon-Mueller, R. B. (1985) 'Women's Work in Third World Agriculture: Concepts and Indicators', ILO Women, Work and Development Series No. 9.

FAO (1984) 'Women in Food Production and Food Security', paper for government consultation on Role of Women in Food Production and Food Security, Harare, Zimbabwe.

Feldman, R. (1981) 'Employment Problems of Rural Women in Kenya', unpublished paper prepared for ILO.

Fortmann, L. (1981) 'The Plight of the Invisible Farmer: The Effect of National Agricultural Policy and Women in Africa', in Dauber and Caine.

Gaobepe, M. G. and Mwenda, A. (1980) 'The Report on the Situation and Needs of Food Supplies, Women in Zambia', Lusaka, FAO.

Goody, J. (1976) *Production and Reproduction*, Cambridge, Cambridge University Press.

Guyer, J. (1981) 'Household and Community in Africa', *African Studies Review*.

Guyer, J. (1983) 'Anthropological Models of African Production: The Naturalisation Problem', Working Paper No. 78, Boston, African Studies Centre.

Hanger, J. and Moris, J. (1973) 'Women and the Household Economy', in Chambers, R. and Moris, J. (eds), *Mwea: An Irrigated Rice Settlement in Kenya*, München, Weltforum Verlag.

IBRD (1979) *Recognising the Invisible in Development: The World Bank Experience*, Washington DC, IBRD.

ICRW (1980) *The Productivity of Women in Development Countries: Measurement Issues and Recommendations*, Washington DC, ICRW.

Jackson, C. (1985) *The Kano River Irrigation Project*, West Hartford, Conn., Kumarian Press.

Kitching, G. (1980) *Class and Economic Change in Kenya: The Making of an African Petite-Bourgeoisie*, New Haven, Conn., Yale University Press.

Mascarenhas, S. and Mnilinyi, M. (1983) *Bibliographic Review of Women in Tanzania*, Uppsala, Uppsala Institute of African Studies.

Moock, P. (1976) 'The Efficiency of Women as Farm Managers: Kenya', *American Journal of Agricultural Economics*, 58.

Muntemba, M. S. (1982) 'Women as Food Producers and Suppliers in the Twentieth Century: The Case of Zambia', *Development Dialogue*, 1–2.

Palmer, I. (1985) 'The Impact of Agrarian Reform on Women', in *Women's Roles and Gender Differences in Development: Case Studies for Planners*, prepared by the Population Council, West Hartford, Conn., Kumarian Press.

Richards, P. (1983) 'Ecological Change and the Politics of African Land Use', *African Studies Review*, 26.

Roberts, P. (1983) 'Feminism in Africa: Feminism and Africa', *Review of African Political Economy*, 27–8.

Safilios-Rothschild, C. (1985) 'The Implications of the Roles of Women in Agriculture in Zambia', New York, Population Council.

Staudt, K. (1978) 'Agricultural Productivity Gaps: A Case Study of Male Preference in Government Policy Implementations', *Development and Change*, 9.

Staudt, K. (1979) 'Women and Participation in Rural Development: Framework for Project Design and Policy Oriented Research', Cadnell University Development Committee, Paper No. 8.

Staudt, K. (1985) 'Agricultural policy implementation: a case study from Western Kenya', in *Women's Roles and Gender Differences in Development: Case Studies for Planners*, prepared by the Population Council, West Hartford, Conn., Kumarian Press.

Tobisson, E. (1984) 'Women, Work, Food and Nutritional Status in Nyamwigura Village, Mara Region, Tanzania', annotated in FAO.

United Nations (1984) 'Improving Concepts and Methods for Statistics and Indicators on the Situation of Women', *Studies in Methods*, Series F, No.33.

Whitehead, A. (1981) ' "I'm hungry, Mum": The Politics of Domestic Budgeting', in K. Young (ed.) *Of Marriage and the Market*, London, CSE Books.

Whitehead, A. (1984) 'Beyond the Household: Gender and Resource Allocation in a Ghanaian Domestic Economy', Paper for Workshop on Conceptualising the Household in Africa, Harvard University, November 1984.

Whitehead, A. (1990a) 'Rural Women and Food Production in Sub-Saharan Africa', in Dreze, J. and Sen, A. (eds), *The Political Economy of Hunger*, Vol. 1, Oxford, Clarendon Press.

Whitehead, A. (1990b) 'Food Crisis and Gender Conflict in the African Countryside', in Bernstein, H. (ed.), *The Food Question: Profits versus People?*, London, Earthscan Publications.

Whitehead, A. (1991), 'Food Production and the Food Crisis in Africa', in T. Wallace and C. March (eds) *Changing perceptions: Writings on Gender and Development*, Oxford, Oxfam.
Youssef, N. and Hetler, C. (1984) 'Rural Households Headed by Women: A Priority Concern for Development', Women and Rural Development Paper No. 31, Geneva, ILO.

4 Employment

Further reading

India

Agarwal, B. (1981) 'Agricultural Modernisation and Third World Women: Pointers for the Literature and an Empirical Analysis', *World Employment Programme Research Paper* No.WEP10/WP21, Geneva, ILO.

Banerjee, N. (1985) *Women Workers in the Unorganized Sector*, Hyderabad, Sangam Books Ltd. (Available from London distributors, 36 Molyneux Street, London W1). This is a detailed and illuminating study of the employment patterns, wage levels, working conditions and social background of 400 women workers in the unorganized sector in Calcutta. Many of these women are employed in the new industries, generally as piece-rate workers.

Baud, I. S. A. (1983) 'Women's Labour in the Indian Textile Industry', *IRIS Reports*, 23, Sweden, Tilbury Institute of Development Research.

Bhatt, E. (1976) *SEWA: Profiles of Self-Employed Women*, Ahmedabad, India.

Billings, M. and Singh, A. (1970) 'Mechanization and the Wheat Revolution: Effects on Female Labour in Punjab', *Economic and Political Weekly*, 5(5), p. 169.

Chatterji, R. (1984) 'Marginalisation of the Induction of Women into Wage Labour: The Case of Indian Agriculture', *WEP Research Working Paper* No.WEP10/WP32, Geneva, ILO. See text for discussion of this article.

Dyson, T. and Moore, M. (1983) 'On Kinship Structure, Female Autonomy and Demographic Behaviour in India', *Population and Development Review*, 9(1).

Government of India (1974) *Towards Equality*, Report of the Committee on the Status of Women in India, New Delhi, Department of Social Welfare. This is an essential source, both of statistical data on women and on trends in women's education, employment and health levels, and on their political participation. It is based on extensive primary and secondary research and contains a valuable summary of key issues and a set of recommendations.

Mencher, J. and Saradamoni, K. (1982) 'Muddy Feet and Dirty Hands: Rice Production and Female Labour', *Economic and Political Weekly*, XVII(52), 25 December, (Review of Agriculture). This is a preliminary report of a study of women in rice production in three different states. It explores interregional differences in the availability of employment, in work patterns and in agricultural practices. It emphasizes the importance of greater disaggregation of workforce data, the effects of women's wages on household viability and the effects of Green Revolution technology on women's employment.

Mies, M. (1980) 'Capitalist Development and Subsistence Production: Rural Women in India', *Bulletin of Concerned Asian Scholars*, XII(1). See text for discussion.

Mies, M. (1982) *The Lacemakers of Narsapur: Indian Housewives Produce for the World Market*, London, Zed Press. This study of women piece-rate workers shows how a 'handicrafts' industry was transformed into a major capitalist export-oriented industry through the use of extremely low-paid, household-based workers whose contribution to the household, the national and the international economy is 'hidden' ideologically in assumptions about their status as non-working housewives.

Sen, G. (1982) 'Women Workers and the Green Revolution' in Beneria, L. (ed.) *Women and Development: The Sexual Division of Labour in Rural Societies*, New York, Praeger Publishers. See text for discussion.

SEWA (1983) (Self-Employed Women's Association) 'We, the Self-Employed Workers', June, Ahmedabad, India.

Sharma, U. (1980) *Women, Work and Property in North West India*, London, Tavistock Publications. This study compares the lives and economic positions of women in two villages in neighbouring states of north-west India. It looks particularly at the roles of different classes of women in agricultural production and relates this to other aspects of their status and to prevailing cultural ideologies such as purdah.

Sharma, U. (1986) *Women's Work, Class and the Urban Household*, London, Tavistock Publications. This study is set in the administrative town of Shimla in north-west India. It examines the interrelationship between women's waged work and their contribution to the maintenance and management of household resources in different social classes.

Standing, H. (1985) 'Women's Employment and the Household: Some Findings from Calcutta', *Economic and Political Weekly*, 20(17), 27 April, (Review of Women's Studies). This is a preliminary account of findings from a study of a random sample of women of the effects on the household of their entry into employment across a range of occupations but with a bias towards the less skilled.

Standing, H. (1991). *Dependence and Autonomy: Women's Employment and the Family in Calcutta*, London/New York, Routledge.

Whitehead, A. (1985) 'Technological Change and Rural Women' in Ahmed, A. (ed.) *Technology and Rural Women: Conceptual and Empirical Issues*, London, Allen & Unwin.

General

Dex, S. (1985) *The Sexual Division of Work: Conceptual Revolutions in the Social Sciences*, Brighton, Wheatsheaf Books.

Heyzer, N. (1986) *Working Women in South-East Asia*, Milton Keynes, Open University Press.

Humphrey, J. (1985) 'Gender, Pay and Skill: Manual Workers in Brazilian

Industry' in Afshar, H. (ed.) *Women, Work and Ideology in the Third World*, London, Tavistock Publications.
Young, K., Wolkowitz, C. and McCullagh, R. (1981) *Of Marriage and the Market*, London, CSE Books.

5 Housing

Further reading

Essential

Blayney, R. and Lycette M. (1983) *Improving the Access of Women-Headed Households to Solanda Housing: A Feasible Down Payment Assistance Scheme*, International Centre for Research on Women, Washington DC.
Chant, C. (1987) 'Household Composition and Housing in Queretaro, Mexico' in Moser, C. and Peake, L. (eds) *Women, Human Settlements and Housing*, London, Tavistock Publications.
Fernando, M. (1987) 'Women's Participation in the Housing Process: The Case of Kirillapone, Sri Lanka' in Moser, C. and Peake, L. (eds) *Women, Human Settlements and Housing*, London, Tavistock Publications.
Fernando, M. (1987) 'New Skills for Women: A Community Development Project in Colombo, Sri Lanka' in Moser, C. and Peake, L. (eds) *Women, Human Settlements and Housing*, London, Tavistock Publications.
Girling, R. H., Lycette, M. and Youssef, N. (1983) *A Preliminary Evaluation of the Panama Self-Help Women's Construction Project*, International Centre for Research on Women, Washington DC.
Kusnir, L. and Largaesdpada, C. (1986) 'Women's Experiences in Self Help Housing Projects' in Bruce, Kohn and Schmink (eds) *Learning about Women and Urban Services in Latin America and the Caribbean*, Population Council, New York.
Lycette, M. and Jaramillo, C. (1984) *Low-Income Housing: A Women's Perspective*, International Centre of Research on Women, Washington DC.
Machado, L. (1987) 'Women and Low-Income Housing in Brazil: Evaluation of the Profilurb Programme in Terms of its Capacity to Define and Reach Female-Headed Households as a Target Group' in Moser, C. and Peake, L. (eds) *Women, Human Settlements and Housing*, London, Tavistock Publications.
Moser, C. (1985b) 'Housing Policy and Women: Towards a Gender Aware Approach', Gender and Planning Working Paper No. 7, Development Planning Unit, University College London.
Moser, C. (1987) 'Residential Struggle and Consciousness; The Experience of Poor Women in Guayaquil, Ecuador' in Moser, C. and Peake, L. (eds) *Women, Human Settlements and Housing*, London, Tavistock Publications.
Moser, C. (1987) 'Women, Human Settlements and Housing: A Conceptual Framework for Analysis and Policy-Making' in Moser, C. and Peake, L. (eds) *Women, Human Settlements and Housing*, London, Tavistock Publications.

Moser, C. (1987) 'Mobilization is Women's Work: Struggles for Infrastructure in Guayaquil, Ecuador' in Moser, C. and Peake, L. (eds) *Women, Human Settlements and Housing*, London, Tavistock Publications.

Moser, C. and Chant, S. (1987) 'The Role of Women in the Execution of Low-Income Housing Projects Training Module' in Moser, C. and Peake, L. (eds) *Women, Human Settlements and Housing*, London, Tavistock Publications.

Moser, C. and Peake, L. (1987) *Women, Human Settlements and Housing*, London, Tavistock Publications.

Nimpuno-Parente, P. (1987) 'Gender Issues in Project Planning and Implementation: The Case of Dandora Site and Service Project, Kenya' in Moser, C. and Peake, L. (eds) *Women, Human Settlements and Housing*, London, Tavistock Publications.

Peake, L. (1987) 'Low Income Women's Participation in the Housing Process: A Case Study from Guyana' in Moser, C. and Peake, L. (eds) *Women, Human Settlements and Housing*, London, Tavistock Publications.

Population Council (1983) *The Performance of Men and Women in the Repayment of Mortgage Loans in Jamaica*, The Population Council (Jamaica Working Group).

Resources for Action (1982a) *Women and Shelter in Honduras*, Washington DC, USAID (Office of Housing).

Resources for Action (1982b) *Women and Shelter in Tunisia: A Survey of the Shelter Needs of Women in Low-Income Areas*, Washington DC, USAID (Office of Housing).

Sara-Lafosse, V. (1984) *Comedoras Comunales: La Mujer Frente a la Crisis, Grupo Trabajo*, Servicios Urbanos y Mujeres de Bajos Ingresos, Lima.

Schmink, M. (1982) 'Women in the Urban Economy in Latin America', Working Paper No.1, Women, Low-Income Households and Urban Services.

Schmink, M. (1984) *The Working Group Approach to Women and Urban Services*, Mimeo, Centre for Latin American Studies, Gainsville, University of Florida.

Sorock, M., Dicker, H., Giraldo, A. and Waltz, S. (1984) *Women and Shelter, Resources for Action, Office of Housing and Urban Programs*, Washington DC, USAID.

Young, K. and Moser, C. (1981) (eds) 'Women and the Informal Sector', *Institute of Development Studies Bulletin*, 12 (3).

Vance, I. (1987) 'Women's Participation in Self-Help Housing: The San Judas Barrio Project, Managua, Nicaragua' in Moser, C. and Peake, L. (eds) *Women, Human Settlements and Housing*, London, Tavistock Publications.

Additional

Agarwal, A. and Anand, A. (1982) 'Ask the Women Who Do the Work', *New Scientist*, 4 November 1982, pp. 302–4.

Caplan, P. (1981) 'Development Policies in Tanzania: Some Implications for Women', *Journal of Development Studies*, 17(3), pp. 98–108.

Chant S. (1984) 'Household Labour and Self-Help Housing in Queretaro,

Mexico', *Boletin de Estudios Latinoamericanos y del Caribe*, 37, pp. 45–68.
Chant, S. (1985b) Family Formation and Family Roles in Queretaro, Mexico, *Bulletin of Latin American Research*, 4(1), pp. 17–32.
International Women's Tribune Centre (IWTC) (1981), 'Women, Money and Credit', *Newsletter* No. 15, New York.
International Women's Tribune Centre (IWTC) (1982), 'Women and Water', *Newsletter* No. 20, New York.
Kindervatter, S. (1983) *Women Working Together for Personal, Economic and Community Development: A Handbook of Activities for Women's Learning and Action Groups*, Washington DC, Overseas Education Fund.
Moser, C. (1982) 'A Home of One's Own: Squatter Housing Strategies in Guayaquil, Ecuador' in Gilbert, A. in association with Hardoy, J. E. and Ramirez, R. (eds) *Urbanisation in Contemporary Latin America: Critical Approaches to the Analysis of Urban Issues*, Chichester, John Wiley, pp. 159–90.
Moser, C. (1989) *Approaches to Community Participation in Urban Development Programmes in Third World Cities*, Economic Development Institute, Washington.
Moser, C. (1989) 'Gender Planning in the Third World: Meeting Practical and Strategic Gender Needs', *World Development*, 7(2).
Moser, C. (1989a) 'Community Participation in Urban Projects in the Third World', *Progress in Planning*, 32, Part 2.
Racelis Hollnsteiner, M. (1982) 'People-Powered Development: Thoughts for Urban Planners, Administrators and Policy Makers', Paper presented at the Regional Congress of Local Authorities for Development of Human Settlements in Asia and the Pacific, 9–16 June 1982, Yoohama, Japan.
Rakodi, C. (1983) 'The World Bank Experience: Mass Community Participation in the Lusaka Upgrading Project' in Moser, C. (ed.) *Evaluating Community Participation in Urban Development Projects*, pp. 18–33.
Schlyter, A. (1984) 'Upgrading Reconsidered – the George Studies in Retrospect', *Bulletin: The National Swedish Institute of Building Research*, M84:4, Lund.
Singh, A. (1980) *Women in Cities: An Invisible Factor in Urban Planning in India*, The Population Council.
UNCHS (1983) 'Community Participation in the Execution of Low-Income Projects in Developing Countries', Nairobi.
UNCHS (1984) *Sites and Services Schemes: The Scope for Community Participation*, Nairobi.

6 Transport

References

Anderson, J. and Panzio, N. (1986) 'Transportation and Public Safety Services that Make Service Use Possible' in Bruce, J. and Kohn, M. (eds) *Learning about Women and Urban Services in Latin America and the Caribbean*, A Report on the Women, Low-Income Households and Urban Services Project of the Population Council.

Dimitriou, H. and Safier, M. (1982) 'A Developmental Approach to Urban Transport Planning', a paper given at the Universities Transport Study Group Seminar 'Transport Planning Research and Education Appropriate to Third World Countries', 21 May 1982, Development Planning Unit, University College London.

Edmonds, G., Goppers, K. and Soderback, M. (1986) 'Men or Machines? An Evaluation of Labour Intensive Public Works in Lesotho', SIDA Evaluation Report, Industry, Lesotho, 4.

Fox, M. (1983) 'Working Women and Travel: The Access of Women to Work and Community Facilities', *APA Journal*, Spring.

GLC Women's Committee (1984a) 'Transport' in *Programme for Action for Women in London*, London, The Greater London Council.

GLC Women's Committee (1984b) 'Detailed Results: Black Afro-Caribbean and Asian Women' in *Women on the Move: GLC Survey on Women and Transport*, Vol. 5, London, The Greater London Council.

GLC Women's Committee (1984c) 'Ideas for Action', ibid., Vol. 6.

GLC Women's Committee (1984d) 'Methodology', ibid., Vol. 7.

GLC Women's Committee (1985b) 'Survey Results: Safety, Harassment and Violence', ibid., Vol. 3.

Hagen, S., Guthrie, P. and Galetshoge, D. (1988) 'LG-34 District Roads: Labour Intensive Improvement and Maintenance Programme Botswana', NORAD Project Review BOT 012.

Hanson, S. and Hanson, P. (1981) 'The Impact of Married Women's Employment on Household Travel Patterns: A Swedish Example', *Transportation*, 10, pp. 165–83.

Kneerim, J. (1980) *Village Women Organise: The Mraru Bus Service*, New York, Seeds.

Molyneux, M. (1985) 'Mobilization without Emancipation? Women's Interests, State and Revolution and Nicaragua', *Feminist Studies*, Vol. 11, No. 2.

Moser, C. (1987a) 'Residential Struggle and Consciousness: The Experience of Poor Women in Guayaquil, Ecuador', in Moser, C. and Peake, L., *Women, Human Settlements and Housing*, London, Tavistock Publications.

Moser, C. (1987b) 'Housing Policy and Women: Towards a Gender Aware Approach', in Moser, C. and Peake, L., *Women, Human Settlements and Housing*, London, Tavistock Publications.

Moser, C. (1989) 'Gender Planning in the Third World: Meeting Practical and Strategic Gender Needs', *World Development*, Vol. 17, No. 11.

Nairobi City Council (1984) Report by the Transport Unit.

Pickup, L. (1984) 'Women's Gender Role and its Influence on Travel Behaviour, *Built Environment*, 10(1), pp. 61–8.

Rosenbloom, S. (1978) 'Editorial: The Need for Study of Women's Travel Issues', *Transportation*, 7, pp. 347–50.

Schmink, M. (1982) 'Women in the Urban Economy in Latin America', Working Paper No. 1, Women, Low-Income Households and Urban Services.

Thomson, J. M. (1983) *Towards Better Urban Transport Planning in Developing Countries*, World Bank Staff Working Papers, No. 600, Washington DC, World Bank.

Further reading

GLC Women's Committee (1985a) 'Survey Results: The Overall Findings', ibid., Vol. 2.

GLC Women's Committee (1985c) 'Initial Research Preliminary to Survey: Women's Discussion Groups', ibid., Vol. 4.

GLC Women's Committee (1986) 'Detailed Results: Differences Between Women's Needs', ibid., Vol. 4.

Lopata, H. Z. (1981) 'The Chicago Women: A Study of Patterns of Mobility and Transportation' in Stimpson, C., Dixler, E., Nelson, M. and Yatrakis, K. (eds) *Women and the American City*, Chicago and London, The University of Chicago, pp. 158–66.

Werkele, G. R. (1981) 'Women in Urban Environment' in Stimpson, C., Dixler, E., Nelson, M. and Yatrakis, K. (eds) *Women and the American City*, Chicago and London, The University of Chicago, pp. 185–211.

World Bank (1982) 'Kenya Rural Access Programme: Case Study', Notes on Women in Development No. 23, Cr. No.651, Office of the Advisor on Women in Development, Washington.

7 Health

References

Andersen, C. and Staugård, F. (1986) *Traditional Midwives – Traditional Medicine in Botswana*, Botswana, Ipelegeng Publishers.

Chauhan, S. K. and Gopalakrishnan, K. (1983) *A Million Villages, A Million Decades*, London, Earthscan.

Harrington, J. A. (1983) 'Nutritional Stress and Economic Responsibility: A Study of Nigerian Women', in Buvinic, M. et al (eds) *Women and Poverty in the Third World*, Baltimore, Johns Hopkins University Press.

Islam, M. (1985) 'Child Wives of Bangladesh', *People*, 12(3), 8–9.

Jørgensen, V. (1983) *Poor Women and Health in Bangladesh: Pregnancy and Health*, Uppsala, SIDA.

Levison, J. F. (1974) *Morinda: An economic analysis of malnutrition among young children in rural India*, Cambridge, Mass.

McLean, E. (1987) 'World Agricultural Policy and its Effect on Women's Health', *Health Care for Women International*, 8(4), 231–8.

Mies, M. (1981) 'The social origins of the sexual division of labour', The Hague, *ISS*, No. 85, January.

Mitra, A. (1985) 'The Nutrition Situation in India', in Biswas, M. and Pinstrup-Andersen, P. (eds), *Nutrition and Development*, Oxford, Oxford University Press.

Moore, M. P. (1973) 'Cross-cultural Surveys of Peasant Family Structures', *American Anthropologist*, 75(13).

Østergaard, L. (1986) 'Women in Development', Seminar Paper, UNDP/Swedish Government Development Policy Seminar for Senior UNDP Staff, 22/9–3/10–1986.

Østergaard, L. (1987) 'The Role of Women as a Motivative and Collective

Force in International Development', *Health Care for Women International*, 8(4), pp. 219–30.
Ratcliffe, J. (1983) 'Social Justice and the Demographic Transition: Lessons from India's Kerala State', in Morley, D. et al., *Practising Health for All*, Oxford, Oxford University Press.
Rifkin, S. B. (1985) *Health Planning and Community Participation: Case Studies in South-East Asia*, London/Sidney/Dover, New Hampshire, Croom Helm.
Rogers, B. (1980) *The Domestication of Women: Discrimination in Developing Societies*, London, Kogan Page.
Royston, E. and Armstrong, S. (eds) (1989) *Preventing Maternal Deaths*, Geneva, WHO.
Sanders, D. and Carver, R. (1985) *The Struggle for Health: Medicine and the Politics of Underdevelopment*, London, Macmillan Education.
Silberschmidt, M. (1986) *Studies on the Local Context of Birth Control in Kisii District, Kenya*, Stockholm, SIDA.
SNDT University (1983) 'Biomass Fuel Hazards for Indian Women', *Bulletin of the Unit on Women's Studies*, May, Bombay.
UNICEF (1982) Programme with and for Women.
Warwick, D. P. (1982) *Bitter Pills: Population Policies and their Implementation in Eight Developing Countries*, Cambridge, Cambridge University Press.
WHO (1978) *Primary Health Care*, Report of the International Conference on Primary Health Care, Alma Ata, USSR.
WHO (1986a) *Maternal Mortality Rates: A Tabulation of Available Information*, pp. 63–87.
WHO (1986b) 'Maternal Mortality: Helping Women off the Road to Death' *WHO Chronicle*, 1986, 40(5).
WHO (1986c) 'A Traditional Practice that Threatens Health – Female Circumcision', *WHO Chronicle*, 40(1).
WHO/UNICEF (1986) *Health Implications of Sex Discrimination in Childhood*, Geneva.
World Bank (1989) 'Investing in people', in *World Bank: Sub-Saharan Africa: From Crisis to Sustainable Growth*, Washington, World Bank.

8 Household management

References

Burfisher, M. E. and Horenstein, N. R. (1985) *Sex Roles in the Nigerian Tiv Farm Household*, Kumarian Press.
Dwyer, D. and Bruce, J. (eds) (1988) *A Home Divided: Women and Income in the Third World*, Stanford University Press.
Feldstein, H. (1986) 'Intrahousehold Dynamics and Farming Systems Research and Extension: Conceptual Framework', Mimeo. Conference paper for Farming Systems Research Conference, Miami.
Mueller, E. (1983) in Buvinic, M. et al. (eds) *Women and Poverty in the Third World*, Baltimore, Johns Hopkins University Press.

Pahl, J. (1983), 'The allocation of money and the structuring of inequalities within marriage', *The Sociological Review.*
Rogers, B. L. (1985) 'Incorporating the Intrahousehold Dimension into Development Projects: A Guide for Planners', Mimeo, Tufts University School of Nutrition.
Roldan, M. (1988) in Dwyer, D. and Bruce, J. (eds) *A Home Divided: Women and Income in the Third World*, Stanford University Press.
Sen, A. (1987) *Gender and Cooperative Conflicts*, World Institute for Development Economics Research. First published 1985 as 'Women, Technology and Sexual Division' in *Trade and Development*, 6, UNCIAD.
Whitehead, A. (1981) in Young, K. et al., *Of Marriage and the Market*, London, CSE Books. Reprinted 1984 London: Routledge & Kegan Paul.

Further Reading

Boulding, E. (1983) in Buvinic, M. et al. (eds) *Women and Poverty in the Third World*, Baltimore, Johns Hopkins University Press.
Buvinic, M., Lycette, M. A. and McGreevey, W. P. (eds) (1983) *Women and Poverty in the Third World*, Baltimore, Johns Hopkins University Press.
Galbraith, J. K. (1973), *Economics and the Public Purpose*, New York, Signet.
Guyer, J. (1979) 'Household Budgets and Women's Income', African Studies Center Working Papers No. 28, Boston University.
ILO (1977) *Employment, Growth and Basic Needs: A One World Problem*, New York, Praeger.
King, E. and Evenson, R. E. (1983) in Buvinic, M., et al. (eds) *Women and Poverty in the Third World*, Baltimore, Johns Hopkins University Press.
WHO/UNICEF (1986) *Health Implications of Sex Discrimination in Childhood*, World Fertility Survey (1983).
Young, K. (1987) 'Benefits and Barriers in the Policy Process', Commonwealth Secretariat Paper for the Second Meeting of Commonwealth Ministers of Women's Affairs.
Young, K. (1988) 'Reflections on Women's Needs' in Young, K. (ed.) *Women and Economic Development*, Berg/UNESCO.
Young, K., McCullough, R. and Wolkowitz, C. (1981) *Of Marriage and the Market*, London, CSE Books. Reprinted 1984, London, Routledge & Kegan Paul.

9 PRACTICAL GUIDELINES

References

AID Program (1990) *A Users Guide to the Office of Women in Development*, US Department of State, Washington DC.
Andersen, C. and Baud, I. (eds) (1987) *Women in Development Cooperation: Europe's Unfinished Business*, EADI Book Series 6.
Commission of the European Communities (1988) 'Women and Development in Community Practice, The Impact of Community Actions on Women', COM VIII/149/82.

Commission of the European Communities (1988) 'Women in Development, Progress Report and Guidelines for an Operational Programme'.

Cornia G., Jolly, R. and Stewart, F. (1987, 1988) *Adjustment with a Human Face*, Vols. 1 and 2, Oxford, Oxford University Press.

Moser, C. (1989a) *Approaches to Community Participation in Urban Development Programmes in Third World Cities*, Economic Development Institute, Washington.

Moser, C. (1989b) 'Gender Planning in the Third World: Meeting Practical and Strategic Gender Needs', *World Development*, Vol. 17, No. 11.

Norwegian Government White Paper No. 96.

OECD (1988) Development Cooperation Report 1987, Paris.

OECD (1988) *Second Monitoring Report on the Implementation of the DAC Guiding Principles to Aid Agencies for Supporting the Role of Women in Development.*

Sara-Lafosse, V. (1984) *Comedores Comunales: La Mujer Frenta A la Crisis*, Lima, Grupo de Trabajo, Servicios Urbanos y Mujeres de Bajos Ingresos.

Schultz, Theodore N. (1980) 'Nobel Lecture: The Economics of being Poor', *Journal of Political Economy*, Vol. 88, No. 4.

United Kingdom Overseas Development Administration (1989) 'Strategy for Implementation of O.D.A.'s Policy on Women in Development'.

Vrouwenberaad Nederlandse Ontwikkelingsinstanties (1988) *Women and Development means Business.*

Further Reading

Most donor countries have published monographs or booklets in their national language and/or in English on gender and development and their policy and/or administrative approach to the subject.

AID (1989) 'A Report to Congress Planning for the Next Decade: A Perspective of Women in Development'.

Australian International Development Assistant Bureau (1990) 'Women in Development Fund, Information and Guidelines'.

Belgian Administration for Development Co-operation and the Royal Academy for Overseas Services (1990) 'Progress for Women in Developing Countries, the Belgian Contribution'.

Organization for Economic Cooperation and Development (1991) 'Third Monitoring Report on the Implementation of the DAC Revised Guiding Principles on Women in Development, 1989', Paris.

United Kingdom, Overseas Development Administration (1988) 'Women, Development and the British Programme, A Progress Report'.

Index